Anthropogenic Soils

Progress in Soil Science

Series Editors:

Alfred E. Hartemink, *Department of Soil Science, FD Hole Soils Lab, University of Wisconsin—Madison, USA*

Alex B. McBratney, *Faculty of Agriculture, Food & Natural Resources, The University of Sydney, Australia*

Aims and Scope

Progress in Soil Science series aims to publish books that contain novel approaches in soil science in its broadest sense – books should focus on true progress in a particular area of the soil science discipline. The scope of the series is to publish books that enhance the understanding of the functioning and diversity of soils in all parts of the globe. The series includes multidisciplinary approaches to soil studies and welcomes contributions of all soil science subdisciplines such as: soil genesis, geography and classification, soil chemistry, soil physics, soil biology, soil mineralogy, soil fertility and plant nutrition, soil and water conservation, pedometrics, digital soil mapping, proximal soil sensing, soils and land use change, global soil change, natural resources and the environment.

More information about this series at http://www.springer.com/series/8746

Jeffrey Howard

Anthropogenic Soils

Springer

Jeffrey Howard
Department of Geology
Wayne State University
Detroit, MI
USA

ISSN 2352-4774 ISSN 2352-4782 (electronic)
Progress in Soil Science
ISBN 978-3-319-54330-7 ISBN 978-3-319-54331-4 (eBook)
DOI 10.1007/978-3-319-54331-4

Library of Congress Control Number: 2017933670

© Springer International Publishing AG 2017
This work is subject to copyright. All rights are reserved by the Publisher, whether the whole or part of the material is concerned, specifically the rights of translation, reprinting, reuse of illustrations, recitation, broadcasting, reproduction on microfilms or in any other physical way, and transmission or information storage and retrieval, electronic adaptation, computer software, or by similar or dissimilar methodology now known or hereafter developed.
The use of general descriptive names, registered names, trademarks, service marks, etc. in this publication does not imply, even in the absence of a specific statement, that such names are exempt from the relevant protective laws and regulations and therefore free for general use.
The publisher, the authors and the editors are safe to assume that the advice and information in this book are believed to be true and accurate at the date of publication. Neither the publisher nor the authors or the editors give a warranty, express or implied, with respect to the material contained herein or for any errors or omissions that may have been made. The publisher remains neutral with regard to jurisdictional claims in published maps and institutional affiliations.

Printed on acid-free paper

This Springer imprint is published by Springer Nature
The registered company is Springer International Publishing AG
The registered company address is: Gewerbestrasse 11, 6330 Cham, Switzerland

While the farmer holds the title to the land, actually it belongs to all the people because civilization itself rests upon the soil

—Thomas Jefferson

To my wife Christy

Preface

Although scientific studies of anthropogenic soils in archaeological and agricultural settings date back to the late nineteenth century, studies of those in urban and mine-related settings mainly date from the late twentieth and early twenty-first centuries. The results of these studies are found in a plethora of different publications from such diverse fields as soil science, archaeology, geology, engineering, environmental science, etc. Hence, the purpose of this book was to collate data from many disparate sources, and organize them into a state-of-the-art compendium of scientific knowledge on the subject of artificial soils. The book focuses on genesis, morphology, and classification; the topics of pollution and reclamation are not addressed.

This book begins with an overview of the historical and geocultural significance of anthropogenic soils, and then examines their relationships with anthropogenic landforms, sediments, soil-forming processes, and artifacts. The classification of anthropogenic soils is discussed next, followed by a systematic look at the major categories of anthropogenic soils associated with agricultural, archaeological, mine-related and urban settings. The basic *modus operandi* was to gather together soil profile descriptions, along with corresponding characterization and analytical data, for each major anthropogenic soil type. Unfortunately, many different systems had been used to make these soil profile descriptions, therefore, using my best judgment, I revamped the original descriptions to conform to the current system being used in Soil Taxonomy by the U.S. Department of Agriculture's Natural Resources Conservation Service. Soil horizons formed in human-transported material are denoted by the carat (^) symbol. However, in order to emphasize their anthropogenic origin, I deviated from Soil Taxonomy and used the asterisk (*) symbol to designate horizons formed in human-altered material. Sometimes there were problems finding analytical data which had been obtained using the same analytical methods. These data are compared and contrasted in tabulated form in this book, but it is recommended that the reader refer to the original sources of data for more quantitative interpretations. Only moist horizon colors are reported unless otherwise indicated.

Soil is a vital resource which is being modified and lost at an accelerating rate on a global scale as a result of urbanization and other human activities. Hence, publication of this book is timely given the extensive and current studies of soil health and resilience, urban soil revitalization, surface mine and other types of land reclamation, contaminated site assessment and remediation, etc. It is also timely given the ongoing lively debate about the postulated Anthropocene Epoch of modern geologic time. It is hoped that the book will be useful to anyone dealing with anthropogenic soils, including urban planners, federal and state environmental protection agencies, environmental consultants and engineers, as well as academicians. Thanks to Alicia Galka for her work with the illustrations, and to friends and colleagues who contributed photographs and other materials for this book: Krysta Ryzewski and Don Adzigian (Department of Anthropology, Wayne State University); Joe Calus, Eric Gano, Debbie Surabian, Luis Hernandez, Richard Shaw, and Shawn McVey (USDA-NRCS); W. Lee Daniels (Department of Soil and Crop Environmental Sciences, Virginia Polytechnic Institute and State University); Russell Losco (Lanchester Soil Consultants); Dr. Richard J. Buckley (University of Leicester Archaeological Services). Special thanks to my mentors Drs. Dan F. Amos and Lucian W. Zelazny, and to my longtime colleague and friend Dr. W. Lee Daniels.

Detroit, Michigan USA
Jeffrey Howard

Contents

1	**The Nature and Significance of Anthropogenic Soils**		1
	1.1 Introduction		1
	1.2 Historical Perspective		4
	1.3 Significance of Anthrosoils		6
	References		7
2	**Geocultural Setting**		11
	2.1 Introduction		11
	2.2 Ancient Civilizations		12
	2.3 Industrialization and Urbanization		14
	2.4 Stratigraphy and Archaeology		16
	References		22
3	**Anthropogenic Landforms and Soil Parent Materials**		25
	3.1 Introduction		26
	3.2 Anthropogenic Landforms		26
		3.2.1 Human Impacts on the Landscape	26
		3.2.2 Natural Versus Anthropogenic Landforms	28
		3.2.3 Physical Landform Classification	29
	3.3 Genetic Landform Classification Systems		31
		3.3.1 NRCS Geomorphic Description System	32
		3.3.2 British Geological Survey Classification	40
	3.4 Anthropogenic Soil Parent Materials		42
		3.4.1 Pedological Description of Parent Materials	42
		3.4.2 Sedimentological Description of Parent Materials	44
	3.5 U.S. Soil Taxonomy		46
		3.5.1 Types of Anthropogenic Parent Materials	46
		3.5.2 Artifacts	47

	3.6	World Reference Base for Soil Resources	49
	3.7	British Geological Survey System	50
	References		50
4	**Human Impacts on Soils**		**53**
	4.1	Introduction	53
	4.2	Anthropedogenic Processes	55
	4.3	Humans as a Soil-Forming Factor	58
	4.4	Rates of Anthropogenic Soil Formation	60
	References		61
5	**Artifacts and Microartifacts in Anthropogenic Soils**		**63**
	5.1	Introduction	64
	5.2	Macroartifacts	67
		5.2.1 Carbonaceous Artifacts	67
		5.2.2 Calcareous Artifacts	71
		5.2.3 Siliceous Artifacts	73
		5.2.4 Ferruginous Artifacts	75
		5.2.5 Miscellaneous Artifacts	77
	5.3	Microartifacts	78
		5.3.1 Optical Characteristics	78
		5.3.2 Specific Gravity	83
		5.3.3 Abrasion pH	86
		5.3.4 Trace Metal Content	87
		5.3.5 Electrical Conductivity	88
		5.3.6 Magnetic Susceptibility	89
	References		89
6	**Classification of Anthropogenic Soils**		**95**
	6.1	Introduction	95
	6.2	U.S. System of Soil Taxonomy	97
	6.3	World Reference Base of Soil Resources	106
	References		112
7	**Anthropogenic Soils in Agricultural Settings**		**115**
	7.1	Introduction	116
	7.2	Types of Anthrosols	118
	7.3	Hortic Anthrosols	119
	7.4	Plaggic Anthrosols	126
	7.5	Terric Anthrosols	132
	7.6	Pretic Anthrosols	133
	7.7	Irragric Anthrosols	138
	7.8	Hydraquic Anthrosols	140
	References		145

8	**Anthropogenic Soils in Archaeological Settings**...............	149
	8.1 Introduction	150
	8.2 Midden Soils	151
	8.3 European Dark Earth Soils.............................	155
	8.4 Cemetery Soils..	161
	8.5 Burial Mound Soils	165
	References...	168
9	**Mine-Related Anthropogenic Soils**..........................	171
	9.1 Introduction	172
	9.2 Geologic Setting.......................................	175
	9.3 Coal Mine-Related Anthrosoils.........................	178
	References...	185
10	**Anthropogenic Soils in Urban Settings**	187
	10.1 Introduction	187
	10.2 Urban Land Classification.............................	190
	10.3 General Characteristics of Urban Soils	192
	10.4 Urban Soils on Residential Land	208
	10.5 Urban Soils on Industrial Land.........................	214
	10.6 Urban Soils Impacted by Coal-Related Wastes..............	219
	10.7 Classification of Urban Soils	223
	References...	225
11	**Epilogue**...	229

Chapter 1
The Nature and Significance of Anthropogenic Soils

Anthropogenic soils (anthrosoils) are soils that have been influenced, modified or created by human activity, in contrast to soils formed by natural processes. They are found worldwide in urban and other human-impacted landscapes. Anthrosoils are formed by: (1) sealing a natural soil beneath pavement or other artificially manufactured impervious material, (2) transformation of a natural soil by human action, or (3) development of a new soil profile in parent materials created and deposited by human action. Anthrosediments (anthropogenic sediments) are sedimentary deposits formed directly and intentionally by an artificial mechanism of sedimentation (e.g., urban fill). The soil profile developed in anthrosediment is an anthrosoil. Geocultural setting and historical context can be used to differentiate four types of anthrosoil: (1) Agricultural, (2) Archaeological, (3) Mine-related, and (4) Urban. Early scientific studies dating back to the late 19th century were focused on anthrosoils in agricultural and archaeological settings, whereas studies of urban and mine-related anthrosoils became common only during the late 20th century. The study of anthrosoils is of fundamental practical importance in the fields of agronomy and archaeology, whereas studies of urban and mine-related soils are often focused on issues of environmental significance.

1.1 Introduction

Soil is an accumulation at the Earth's surface of organic matter, and the mineral by-products of weathering. Multiple natural processes generally operate simultaneously to form soil. The relative intensities of these processes are a function of five soil-forming factors (Jenny 1941): (1) climate, (2) landscape position (slope), (3) organisms, (4) parent material, and (5) time. In the case of anthropogenic soils, one or more of these factors may be altered by human activities. Indeed, humans have been regarded by some as a sixth soil-forming factor (Effland and Pouyat 1997; Dudal 2005; Leguedois et al. 2016).

Soil is usually a natural body characterized by horizons (layers) that are distinguishable from the initial parent material by additions, losses, transfers and transformations of energy and matter or, in the absence of horizons, by the ability to support rooted plants in a natural environment. The upper limit of soil is the boundary between soil and either air, shallow water, live plants, or plant materials that have not yet begun to decompose. In Soil Taxonomy, areas are not considered to have soil if the surface is permanently covered with water too deep (>2.5 m) for the growth of rooted plants. The lower limit of soil is at the boundary with unweathered parent material, or at 200 cm, whichever is shallower (Soil Survey Staff 2014).

Anthropogenic soils, also known as technogenic or man-made soils, are those that have been influenced, modified or created by human activity, as opposed to natural or native soils, which are formed by natural processes. They are found worldwide in urban and other human-influenced landscapes. Anthropogenic soils may have been so modified by human activity that the original soil remains only as a buried profile, or has been so drastically changed that it is no longer recognizable. They typically have characteristics that lie beyond the range of natural soil types, and require classification as a different type of soil.

Anthrosoils (anthropogenic soils) are formed by three basic mechanisms (Fig. 1.1): (1) sealing a natural soil beneath one or more layers of artificially manufactured impervious material (e.g., concrete pavement), (2) transformation of a natural soil by human action (metagenetic), or (3) development of a new soil profile in parent materials (anthrosediments) created and deposited by human action (Fig. 1.2), usually with the aid of mechanized equipment (neogenetic). Each type can be modified subsequently by reclamation methods used with the goal of

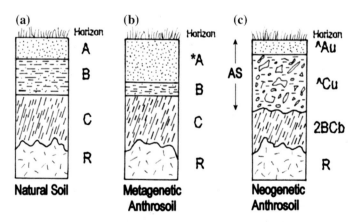

Fig. 1.1 Mechanisms of anthropogenic soil formation: **a** natural soil profile formed from granite bedrock; **b** natural soil changed into anthrosoil by long-term additions of organic matter (manuring) and deep plowing (metagenetic anthrosoil); **c** new soil profile developed in an anthrosedimentary deposit of urban fill containing demolition debris (neogenetic anthrosoil). *AS* Anthrosediment

1.1 Introduction

Fig. 1.2 Profile of neogenetic anthrosoil at former demolition site: **a** previous site was demolished before construction of building (derelict) in 1942; **b** anthropogenic soil profile developed in fill overlies buried natural soil. Photos by J. Howard

revitalizing the soil. Anthrosoils can have the combined features of natural soils and other specific properties developed in a human-impacted environment. Strictly speaking, any natural soil buried by an anthropogenic deposit is human-impacted regardless of the thickness of the artificial capping. However, for interpretive and mapping purposes, anthropogenic soils have been defined in various classifications systems as having developed in an artificial capping ≥ 30–50 cm thick.

Anthrosediments (anthropogenic sediments) are sedimentary deposits formed directly and intentionally by an artificial mechanism of sedimentation. For example, when soil and other earth materials are excavated, moved around, and backfilled along with building demolition debris, an anthropogenic sedimentary deposit (anthrosediment) is formed. The soil profile developed in these artificial parent materials is an anthrosoil. Human-deposited parent material may be comprised of natural materials (mine spoils, excavated soil, rock or regolith, etc.), artificial materials (brick, mortar, concrete, slag, etc.), or a mixture of both. In any case, this parent material can be regarded as a new anthropogenic sedimentary deposit if it is created directly and intentionally by human action. Natural sedimentation rates can be increased or reduced by human activity. For example, overgrazing by farm animals or building construction in a drainage basin can cause accelerated soil erosion which results in an increase in the volume of fluvial sediment downstream. In contrast, dam building may dramatically reduce sediment volume in a drainage basin downstream from the dam. These sedimentary deposits are *not* anthrosediments because the mechanism of sedimentation is natural, i.e. only the rate of deposition is affected indirectly by human action. It is also not generally possible in the field to distinguish such sediments from those of purely natural origin. To facilitate discussion, soils and sediments of anthropogenic origin will be referred to herein as anthrosoils and anthrosediments, respectively.

Four distinct categories of anthrosoil are distinguished in this book, based on geocultural setting and historical context: (1) Agricultural, (2) Archaeological, (3) Mine-related, and (4) Urban. In agricultural settings, anthrosoils are typically formed as a result of human modifications of a pre-existing natural soil by organic or inorganic materials additions over long periods of time in order to enhance soil fertility and crop productivity. They are also formed in rice paddies, and as a result of materials added as a result of irrigation. Anthrosoils are characteristic of archaeological sites of human occupation, and where humans have impacted soils through horticultural, ceremonial and burial activities. Mine-related anthrosoils are primarily associated with modern landscapes created by the surface mining of coal. In urban settings, anthrosoils are found in association with residential, industrial and manufactured land. Urban and mine-related anthrosoils are often highly compacted by earthmoving equipment, and are sometimes formed in parent materials that were imported from offsite. Urban, archaeological, and certain agricultural anthrosoils are unique in that they often contain artifacts (objects of anthropogenic origin), sometimes in great abundance, which may be comprised of substances not found in nature. Anthrosoils are also associated with suburban settings, streets, highways, and utility line or pipeline construction areas.

1.2 Historical Perspective

The written record of early knowledge about soils can be traced back about 6000 years to the ancient civilization of Mesopotamia. This record, along with early accounts from China, refers to agricultural properties and problems associated with soils. Early systems of soil classification dating from about 4000 B.P. were used in ancient China and Greece to assess agricultural productivity for land taxation. The Chinese were also managing soil fertility at least 2000 years ago (Brevik and Hartemink 2010). In Europe, non-scientific knowledge about the impacts of organic matter and fertilizer on soil and plant nutrition dates back to the 16th century or earlier. The first quasi-scientific studies of anthrosoils and anthrosediments were probably those carried out during the 18th century as part of archaeological investigations of well known European sites such as Stonehenge and Herculaneum. The first such investigation in the United States was in 1784 when Thomas Jefferson excavated a Native American burial mound in Virginia. Early scientific (stratigraphic) excavations were being done by the 1870s and 1880s, when sites such as Troy and the Great Pyramid were studied. Modern archaeological studies of many Roman sites in England were being carried out during the late 19th and early 20th century (Fig. 1.3), and by the 1920s and 1930s at Clovis Paleoindian sites in the United States. Similarly, the first scientific studies of the impacts of fertilizers on soils were being carried out in Europe by the mid-1800s, and Dokuchaev and Darwin were carrying out early pedological studies by the 1870s and 1880s. In the United States, the effects of humans on soils were certainly being studied in earnest by the 1930s as a result of the Dust Bowl, particularly after the establishment of the Soil Conservation Service in 1935.

The need for better soil survey and geological information in urban areas was recognized by the 1950s and 1960s in the United States (Robinson et al. 1955; McGill 1964; Wayne 1969). At that time, soils in urbanized terrain were generally

1.2 Historical Perspective

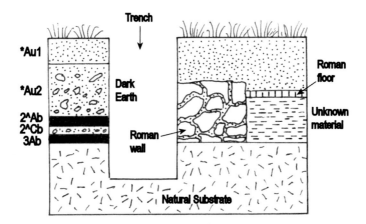

Fig. 1.3 Tentative re-interpretation of one of the earliest known scientific descriptions of an anthropogenic soil at a Roman period archaeological site in Abinger, England investigated by Darwin (1881). He recognized that the "vegetable mould" (topsoil, *Au1*) had formed as a result of earthworm casting activity. Another anthropogenic ^A horizon may have formed in human-transported material created during Roman times (2^Ab), which buried the A horizon of the original natural soil (3Ab)

mapped simply as "made-land" or "urban-land" on soil survey maps. Papers addressing the importance of man as a factor of soil formation began to appear during the 1960s (Bidwell and Hole 1965; Yaalon and Yaron 1966), and several papers and books on urban geology and soils were published during the 1970s (Leggett 1973, 1974; Miller 1978). The first detailed studies of anthrosoils formed in mine spoils were also initiated during the 1970s (e.g., Sencindiver 1977; Howard 1979; Shafer 1979), and the term Anthrosol was proposed as a new soil order (Kosse 1980). Difficulties in applying the concepts of the newly developed system of Soil Taxonomy (Soil Survey Staff 1975) led to the first systematic study of anthrosoils in Washington, D.C. by the Soil Conservation Service during the 1980s (Short et al. 1986a, b). Soon thereafter, in 1988, the International Committee on Anthropogenic Soils (ICOMANTH) was convened with the goal of formulating a classification system for anthrosoils to be used by the Soil Conservation Service. It would be 26 years before the final recommendations of ICOMANTH were formally adopted (Soil Survey Staff 2014).

Scientific publications dealing with anthrosoils became more common during the 1990s. Several books devoted to the topic appeared (e.g., Bullock and Gregory 1991; Craul 1992, 1999), and the issue of human impacts on pedogenesis began to be seriously addressed (Amundson and Jenny 1991; Effland and Pouyat 1997). However, soil survey maps continued to show urban areas as "made-land," and geological maps of urban areas generally ignored the presence of anthropogenic surficial deposits. The first known U. S. Geological Survey map to show anthropogenic fill was that of Washington, D.C. (Fleming et al. 1994), and a USGS program was initiated to develop urban geologic maps of greater societal value

(Bernknopf et al. 1996). The number of publications dealing with anthrosoils swelled during the 2000s (Capra et al. 2015), and the term "Anthropocene" was introduced to characterize the current epoch of anthropogenic global change (Crutzen 2002). This led to an extensive and ongoing discussion by geologists about whether the stratigraphic code should be amended to include the Anthropocene as a formal Epoch of the Quaternary Period (e.g., Zalasiewicz et al. 2010; Certini and Scalenghe 2011; Waters et al. 2014; Howard 2014). A scientific classification of made-land was proposed (Rosenbaum et al. 2003), and the term Technosol was introduced (Rossiter and Burghardt 2003). The first National Cooperative Soil Survey map appeared which included anthropogenic soil series (New York City Soil Survey Staff 2005), and Anthrosol and Technosol soil orders were formally incorporated into an international system of soil classification known as the World Reference Base (IUSS Working Group 2006). Extensive numbers of papers dealing with many aspects of anthrosoils have been published since 2010 (Capra et al. 2015).

1.3 Significance of Anthrosoils

Anthrosoils often directly impact urban redevelopment, hence they are of significance to urban planners, water and other utilities, public health agencies, state and county transportation departments, federal and state environmental protection agencies, builders and developers, insurance companies, environmental consultants, engineering firms, and others. Anthrosoils are also commonly associated with surface and groundwater contamination, especially in urban areas. Many urban areas worldwide have large tracts of vacant land created by building demolition. The characteristics of demolition site soils are important because this open space has attracted great interest in terms of repurposing vacant land for brownfield redevelopment, urban agriculture (Fig. 1.4), green infrastructure, and perhaps as a potential repository for excessive combined storm water overflow. The global population explosion of the last 100 years has drastically reduced the amount of arable land per person. Hence, reclamation and revitalization of anthrosoils on urban and strip mined land may therefore be necessary for future agricultural production simply to sustain the world's burgeoning population. The importance of human impacts on agricultural soils in rural areas has been recognized for many years. There are many serious environmental issues such as soil erosion, loss of soil fertility and resilience, and adverse outcomes related to fertilizer runoff and land disposal of agricultural or urban wastes.

A clear understanding of anthropogenic soil properties is needed by archaeologists to determine if artifacts are in their original context, and whether or not features seen in soil are artificial or natural. Anthrosoil properties can also impact the preservation of buried artifacts in urban areas (Nord et al. 2005; Howard et al. 2015). Soil is the largest terrestrial pool of carbon and is therefore a critical component of the carbon cycle. Hence, anthrosoils may bear on the problem of global

1.3 Significance of Anthrosoils

Fig. 1.4 The rising popularity of urban gardening has drawn much attention to the study of anthropogenic soils. This garden is located in Detroit, Michigan USA where nearly a third of the city is now vacant land created by building demolition. Photo by J. Howard

warming because carbonate formation in urban soils is thought to be one method to capture and sequester excess atmospheric carbon dioxide (Renforth et al. 2009; Washbourne et al. 2012). According to the World Health Organization, antibiotic resistance is a top health issue worldwide. Two million Americans are infected annually with diseases resistant to antibiotics, and as many as 15,000 die. Antibiotic synthesis originally evolved from soil microbes, and over 80% of antibiotics in clinical use today were derived from soil bacteria. There is growing evidence for exchanges of antibiotic-resistant genes between environmental and pathogenic bacteria (Knapp et al. 2011; Seiler and Berendonk 2012). Hence, antibiotic-resistant bacteria in agricultural and other anthrosoils represent a food safety issue of growing concern to the medical community.

References

Amundson R, Jenny H (1991) The place of humans in the state factor theory of ecosystems and their soils. Soil Sci 151:99–109

Bernknopf RL, Brookshire DS, Soller DR, McKee M.J, Sutter JF, Matti JC, Campbell RH (1996) Societal value of geologic maps, vol 1111. U. S. Geological Survey Circular, 53 pp

Bidwell OW, Hole FD (1965) Man as a factor of soil formation. Soil Sci 99:65–72

Brevik EC, Hartemink AE (2010) Early soil knowledge and the birth and development of soil science. Catena 83:23–33

Bullock P, Gregory PJ (1991) Soils in the urban environment. Blackwell Sci. Pub., Oxford, England 174 pp

Capra GF, Ganga A, Grilli E, Vacca S, Buondonno A (2015) A review of anthropogenic soils from a worldwide perspective. J Soils Seds 15:1602–1618

Certini G, Scalenghe R (2011) Anthropogenic soils are the golden spikes for the anthropocene. Holocene 21:1267–1274

Craul PJ (1992) Urban soil in landscape design. Wiley, New York 396 pp

Craul PJ (1999) Urban soils—applications and practices. Wiley, New York 366 pp

Crutzen PJ (2002) Geology of mankind. Nature 415:23

Darwin C (1881) The formation of vegetable mould through the action of worms with observations on their habits. John Murray, London

Dudal R (2005) The sixth factor of soil formation. Eurasian Soil Sci 38:S60–S65

Effland WR, Pouyat RV (1997) The genesis, classification, and mapping of soils in urban areas. Urban Ecosyst 1:217–228

Fleming AH, Drake AA, McCartan L (1994) Geologic map of the Washington West quadrangle, District of Columbia, Montgomery and Prince George's Counties, Maryland, and Arlington and Fairfax Counties, Virginia. USGS Geological quadrangle map GQ-1748, Scale 1:24,000

Howard JL (1979) Physical, chemical and mineralogical properties of mine spoils derived from the Pennsylvanian Wise Formation, Buchanan County, Virginia. M. S. thesis, Department of Agronomy, Virginia Polytechnic Institute and State University, Blacksburg, VA, 109 pp

Howard JL (2014) Proposal to add anthrostratigraphic and technostratigraphic units to the stratigraphic code for classification of holocene deposits. The Holocene 24:1856–1861

Howard JL, Ryzewski K, Dubay BR, Killion TW (2015) Artifact preservation and post-depositional site-formation processes in an urban setting: a geoarchaeological study of a 19th century neighborhood in Detroit, Michigan, USA. J Archaeol Sci 53:178–189

IUSS (International Union of Soil Science) Working Group (2006) World reference base for soil resources 2006. World Soil Resources Report 103, Food and Agriculture Organization United Nations, Rome, Italy, 145 p

Jenny H (1941) Factors of soil formation. McGraw-Hill, New York

Knapp CW, McCluskey, SM, Singh BK, Campbell CD, Hudson, G, Graham DW (2011) Antibiotic resistance gene abundances correlate with metal and geochemical conditions in archived Scottish soils. PLoS ONE:6

Kosse AD (1980) Anthrosols: proposals for a new soil order. Agron. Abs., ASA, CSSA, SSSA, Madison, WI, p 182

Legget RF (1973) Cities and geology. McGraw-Hill, New York 624 pp

Legget RF (1974) Engineering-geological maps for urban development. In: Ferguson HF (ed) Geologic mapping for environmental purposes. Engineering Geology Case Histories no. 10, Geological Society of America, Boulder, pp 19–21

Leguedois S, Sere G, Auclerc A, Cortet J, Huot H, Ouvrard S, Watteau F, Schwartz C, Morel JL (2016) Modeling pedogenesis of technosols. Geoderma 262:199–212

McGill JT (1964) Growing importance of urban geology, vol 487. USGS Circular, 4 pp

Miller FP (1978) Soil survey under pressure: the Maryland experience. J Soil Water Conserv 33:104–111

New York City Soil Survey Staff (2005) New York city reconnaissance soil survey. United States Department of Agriculture, Natural Resources Conservation Service, Staten Island, NY, 52 pp

Nord AG, Tronner K, Mattsson E, Borg GC, Ullen I (2005) Environmental threats to buried archaeological remains. Ambio 34:256–262

Renforth P, Manning DAC, Lopez-Capel E (2009) Carbonate precipitation in artificial soils as a sink for atmospheric carbon dioxide. Appl Geochem 24:1757–1764

Robinson GH, Porter HC, Obenshain SS (1955) The use of soil survey information in an area of rapid urban development. Soil Sci Soc Am Proc 19:502–504

Rosenbaum MS, McMillan AA, Powell JH, Cooper AH, Culshaw MG, Northmore KJ (2003) Classification of artificial (man-made) ground. Eng Geol 69:399–409

Rossiter G, Burghardt W (2003) Classification of urban and industrial soils in the world reference base for soil resources. J Soils Seds 7:96–100

References

Seiler C, Berendonk TU (2012) Heavy metal driven co-selection of antibiotic resistance in soil and water bodies impacted by agriculture and aquaculture. Front Microbiol 3 (10 pp)

Sencindiver JC (1977) Classification and genesis of minesoils. Ph. D. Dissertation, West Virginia University, Morgantown, West Virginia

Shafer WM (1979) Variability of mine soils and natural soils in Southeastern Montana. Soil Sci Soc Am J 43:1207–1212

Short JR, Fanning DS, McIntosh MS, Foss JE, Patterson JC (1986a) Soils of the mall in Washington, D.C. I: statistical summary of soil properties. Soil Sci Soc Am J 50:699–705

Short JR, Fanning DS, McIntosh MS, Foss JE, Patterson JC (1986b) Soils of the mall in Washington, D.C. II: genesis classification and mapping. Soil Sci Soc Am J 50:705–710

Soil Survey Soil Staff (1975) Soil taxonomy—a basic system of soil classification for making and interpreting soil surveys. U. S. Dept. Agric., Agricultural handbook 436, Washington, DC

Soil Survey Staff (2014) Keys to soil taxonomy, 12th edn. U.S. Department of Agriculture, Natural Resources Conservation Service (372 pp)

Washbourne CL, Renforth P, Manning DAC (2012) Investigating carbonate precipitation in urban soils as a method for the capture and storage of atmospheric carbon dioxide. Sci Total Envir 431:166–175

Waters CN, Zalasiewicz JA, Williams M, Ellis MA, Snelling A (2014) A stratigraphical basis for the Anthropocene. In: Waters CW, Zalasiewicz JA, Williams M, Ellis M, Snelling A (eds) A stratigraphical basis for the anthropocene, vol 395. Geological Society of London Special Publication, pp 1–21

Wayne WJ (1969) Urban geology—a need and a challenge. Proc Indiana Acad Sci 78:49–64

Yaalon DH, Yaron B (1966) Framework for man-made soil changes—an outline of metapedogenesis. Soil Sci 102:272–277

Zalasiewicz J, Williams M, Steffen M, Crutzen P (2010) The new world of the anthropocene. Environ Sci Technol 44:2228–2231

Chapter 2
Geocultural Setting

Since ancient times, major impacts of human culture on the landscape have been generally associated with the growth of agriculture and cities. Hence, the nature and global extent of anthropogenic soils is linked to the sociological and geographical aspects of civilization (geocultural setting). The history of anthrosoils can be traced back to the early days of agriculture in the "fertile crescent," and the first city-based civilization in ancient Mesopotamia about 5000 years ago. Here some of the first anthrosoils were formed as a consequence of long-term horticultural activities and salinization resulting from irrigation. Anthropogenic soils were widely produced as a result of agriculture-induced soil degradation, and partly contributed to the demise of ancient civilizations. The rate of anthrosoil formation began to increase rapidly on a global scale during the 18th century when the Industrial Revolution initiated an ongoing period of exponential population growth and urban expansion. The rates have increased even faster since the end of WWII as a consequence of globalized economies, and the explosive growth in population, industrialization and urbanization. The rate of urban soil formation is increasing as a result of the shifting of populations from rural to urban areas. The current human-impacted chapter of Earth history has come to be known as the Anthropocene Epoch. The appearance of anthropogenic soils in the stratigraphic record has been proposed as a possible "golden spike" for defining the onset of the Anthropocene.

2.1 Introduction

The nature and extent of anthropogenic soils depends to a large extent on geocultural setting, i.e. the geographical and sociological aspects of a form or stage of civilization. Since ancient times, major impacts of human culture on the landscape have been generally associated with agriculture and cities. Hence, the creation of significant areas of anthrosoils and anthrosediments can be traced back to the rise of civilization in Mesopotamia about 5000 years ago. The impacts of urbanization on soils remained relatively localized and limited until the 18th century when the Industrial Revolution initiated an ongoing period of exponential global population growth and urban expansion. The rates of population growth, industrialization and urbanization have increased even faster since the end of WWII, along with a growing trend of shifting population from rural to urban areas. The global impact of industrialization and urbanization has become so profound that there is now a proposal being debated that the geologic time scale be amended to include an "Anthropocene Epoch," corresponding to the current human-impacted chapter of Earth history.

2.2 Ancient Civilizations

Agriculture originated during the early Holocene about 9000 B.C., and there is evidence for early irrigation by ~7500 B.C. (Brevik and Hartemink 2010). The first urbanized, state-level societies based on cities were present in Mesopotamia and Egypt by about 3200 B.C. (Fagan 2011). A city is a central, relatively densely populated settlement, which provides services for surrounding agricultural villages in a given region. The city is dependent on these villages, and typically serves as a marketplace and trading center for food and other goods. A city has a population of at least several thousand, and more complex social and economic organization than that of small farming communities. The earliest cities were small (<1000 ha in size), often surrounded by a protective wall (Childe 1950), and had no more than 5000–15,000 inhabitants. Their size was limited by agricultural production because 50–90 farmers were probably required to support one city-dweller (Davis 1955).

Although the reasons for the development of city-based civilizations are still being debated, it seems clear that ancient cities usually grew from small settlements founded in specific locations as a result of local geological conditions. Key geological features included soil fertility, surface water navigation, such as river crossings and ocean ports located along coasts or at the mouth of rivers, drinking water supply, natural military defense, supply of suitable building materials, and mineral deposits (Legget 1973). Whether or not it was the principle factor, soil fertility obviously played a major role in the origin and growth of cities. Early settlements were often sited on alluvial floodplains because of soil fertility. Indeed, the "cradle of civilization" is also known as the "fertile crescent," a region extending along the valleys of the Tigris-Euphrates and Nile Rivers. Irrigation agriculture, large food surpluses, and a diversified farming economy, were also probably important factors contributing to the origin of civilization (Fagan 2011). Thus, the impacts of ancient human culture on the landscape were determined partly by the original locations chosen for the founding of cities.

It is well known that many civilizations have risen and fallen over the past 5000 years. Although each civilization has a unique story, with multiple factors influencing its demise, soil degradation and other detrimental human-impacts on the landscape definitely help explain why early civilizations collapsed. Ancient peoples abused their soils and environment, and their civilizations declined as a result of adverse ecological consequences. The fact that ancient populations were often greater than those of modern times shows that some civilizations never recovered. Mayan soils still have not recovered, even after more than 1000 years of abandonment to the rainforest (Olson 1981).

Early cities were able to grow because of the domestication of cattle and other livestock, and the contemporaneous development of irrigation and the plow about 5500 B.C. This greatly increased agricultural productivity and reduced the number of farmers needed to support each city resident (Montgomery 2012). The Sumerian city of Uruk is thought to have had a population of about 50,000 by 3000 B.C., and at its peak about 1800 B.C., the Mesopotamia Empire may have had a population of 15–20 million. Similarly, the population of the Mayan civilization grew from 200,000 in 600 B.C., to perhaps 6 million in A.D. 800 (Lowdermilk 1953). Unfortunately, agriculture in fertile valley bottoms allowed populations to grow excessively. Once all of the available space for farming on bottomlands was used up, it became necessary to plant crops and carry out animal husbandry on sloping land. Overgrazing by sheep and goats reduced or eliminated vegetal cover, trampling of soil by livestock reduced infiltration, deforestation, and plowing across the contour all promoted gullying. Continuous exposure of bare soil to rainfall and surface runoff caused geologically rapid erosion of hillslope soils. The ancient histories of the Middle East and China are replete with cases where soils were eroded to bedrock, and buildings downstream were partially buried by the resulting anthropogenic sediment (Lowdermilk 1953). The collapse of the Mesopotamian civilization is thought to have occurred largely because eroded soil produced silt that plugged-up irrigation systems. Soil erosion can be dramatically rapid, or so slow that farmers do not perceive that it is happening. In ancient times, crop failures also resulted from salinization caused by irrigation of salty dry climate soils, and by depletion of nutrients and organic matter (Artzy et al. 1988; Montgomery 2007, 2012).

Early civilizations typically had cities with populations numbering in the tens of thousands, but cities with populations >100,000 began to appear during the Iron Age after about 1300 B.C. This population growth is attributable partly to the development of iron plows, which further increased agricultural productivity. Alexandria, Egypt probably had a population in excess of 600,000 by 200 B.C., and Rome, Italy had 1,000,000 inhabitants by about A.D. 1. Although many early civilizations recognized differences in soils and adjusted their cropping patterns based on differences in soil fertility, the ancient Greek, Roman, Chinese, and Mayan civilizations had developed relatively advanced agricultural methods by about 500 B.C., and this development continued into the Middle Ages (Brevik and Hartmink 2010). Thus, major human impacts on deforestation and soil erosion were widespread by 1000 B.C., particularly in China, and accelerated after A.D. 1000. By the sixteenth century, soil degradation was out of control throughout the civilized world as more advanced societies began engineering their environments.

2.3 Industrialization and Urbanization

In modern times, anthrosoils are not only primarily a consequence of agriculture but also urbanization, which is the process by which towns and cities are established, and become larger through construction of buildings, streets, railroads, and other artificial infrastructure. Urbanization has been increasing exponentially as a result of the global population explosion. This is being fueled by industrialization, a process that transforms an agrarian society into one with an economy based on manufacturing. The Industrial Revolution began in Britain about A.D. 1750, and had spread throughout Europe and the United States by 1900. It was based initially on the use of fossil fuels and the invention of steam-powered machinery. The rates of population growth, industrialization and urbanization increased sharply after WWII in what has been called "the Great Acceleration" (Steffen et al. 2011). This phenomenon has been further stimulated by globalization, i.e. the global exchange of cultures, products and services, and the development of an international network of economic systems. The global integration of economic and cultural activities can be traced back to the 19th century, or earlier, as a result of technological advances in various forms of transportation. However, the global development of interdependent economic systems was greatly enhanced by the reduction of international trade barriers, and the advent of electronic communication, particularly mobile phones and the Internet, which connected billions of people in new ways, beginning around 2000.

At the start of the Christian Era ~A.D. 1, the total population of the world was about 250 million. Global population increased three-fold in seven centuries reaching 700 million by 1750. As a result of the Industrial Revolution global population increased from 1 billion in 1800, to 3 billion in 1950. During the Great Acceleration, world population doubled in just 50 years reaching 6 billion by 2000. Global population reached 7.6 billion in 2016, and is projected to reach 9.6 billion by 2050. The magnitude of this growth rate can be appreciated by considering the fact that in A.D. 1800 no more than 50 cities worldwide had a population >100,000. Today, there are thousands of cities with a population >100,000, and hundreds with populations >1,000,000. Modern cities often consist of a vast metropolitan area, i.e. a central city together with its suburbs. There are also many regions comprised of a number of cities, large towns, and other urban areas that have merged to form one continuous urban and industrially developed area known as a conurbation or urban agglomeration. The conurbations of Osaka, Karachi, Jakarta, Mumbai, Shanghai, Manila, Seoul, and Beijing in Asia each have populations over 20 million people, whereas Delhi and Tokyo are each projected to approach or exceed 40 million people within the coming decade. Elsewhere, Mexico City, São Paulo, New York, Lagos, Los Angeles, and Cairo will each soon have a population of more than 20 million people.

Another factor that has influenced the origin and extent of anthrosoils is the move of populations from rural to urban areas. In A.D. 1950, only 20% of the world population lived in cities. By 2008, half the global population lived in urban areas,

2.3 Industrialization and Urbanization

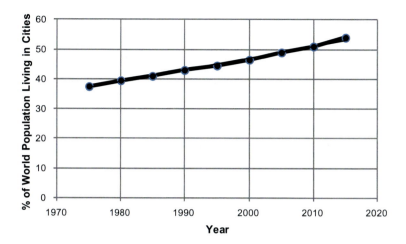

Fig. 2.1 Proportion of the world population living in urban areas. After United Nations (2014)

and the United Nations projects that by 2050, 86% of the developed world's population will be urbanized (Fig. 2.1). Although city size is not necessarily directly related to population size, because of variations in population density, the amount of urban land worldwide has increased dramatically with global urban population growth. The amount of urban land in the U.S. quadrupled between 1945 and 2002 (Lubowski et al. 2006), and urban land is expected to triple in size worldwide between 2000 and 2030 (Seto et al. 2012).

The 20th century saw the invention of gasoline- and diesel-powered engines, and the explosive growth of the automobile industry beginning in the 1920s. These inventions also led to the development and proliferation of mechanized earth-moving and farm equipment, especially during the Great Acceleration following WWII. Mechanized farm equipment, organochlorine pesticides, crop genetics and other agricultural developments increased crop production to the point that even a relatively small number of farmers could support vast numbers of urban residents. Mechanized equipment used for earth moving, mining, terracing, and logging led to global signals in increased sediment discharge in most large rivers. However, by the 1950s, sediment discharge had decreased for most major rivers as a result of dam building. These collective impacts caused major changes in terrestrial sediment transport, and in sediment volumes discharged into the oceans (Syvitski et al. 2011). By the 1990s, it was recognized that the amount of soil and rock moved by humans was equal to, or greater than, that of natural processes (Hooke 1994). The rate at which humans are moving earth is increasing exponentially in response to global population growth (Hooke 2000; Wilkinson and McElroy 2007). It is now estimated that 50% or more of the Earth's surface has been modified by human activities such as agriculture, mining and urbanization (Hooke et al. 2012). Urban renewal has further increased the formation of urban anthrosoils especially through building demolition. As major cities age, there is a need for rebuilding infrastructure

Fig. 2.2 Aging infrastructure is common in major urban areas worldwide: **a** building demolition during urban renewal; **b** demolition debris typical of that found in modern urban soils (wood, brick, mortar, concrete, cinder block, etc.). Photos by J. Howard

and repurposing vacant urban land created by building demolition (Fig. 2.2). Large tracts of vacant urban land are now found in major cities worldwide.

2.4 Stratigraphy and Archaeology

Stratigraphy is the scientific study of strata (layers of rock and sediment) for the purpose of formally naming and classifying them in terms of the geologic time scale, i.e. the global system of organizing strata according to intervals of geologic time. Traditionally, the youngest interval of geologic time was the Holocene Epoch, which began 10,000 ^{14}C years before present (A.D. 1950), or 11,477 ± 85 calendar years ago (Orndorff 2007). However, the term "Anthropocene" was introduced recently to describe the current human-impacted epoch of Earth history (Crutzer and Stoermer 2000; Crutzen 2002). Subsequently, there has been a growing discussion by geologists about whether the stratigraphic code should be amended to include the Anthropocene as a formal Epoch of the Quaternary Period and, if so, how it should be defined (e.g., Zalasiewicz et al. 2010, 2011; Gale and Hoare 2012). Many different indicators have been cited as possibly marking the onset of the Anthropocene Epoch including the extinction of Pleistocene megafauna, the advent of agriculture, the domestication of plants and animals, shell middens, changes in the gas composition of Earth's atmosphere, airborne deposition of coal combustion products, radioactive fallout from nuclear explosions, changes in sediment production, artificial ground, and so forth (e.g., Crutzen and Stoermer 2000; Ruddiman 2007; Syvitski et al. 2011; Price et al. 2011; Erlandson 2013; Smith and Zeder 2013; Oldfield 2015; Waters et al. 2016). It has also been suggested that the first appearance of anthropogenic soils in the geologic record be used as a "golden spike" for defining the base of the Anthropocene (Certini and Scalenghe 2011).

2.4 Stratigraphy and Archaeology

Two basic opposing views have emerged, which can be referred to as the late versus early Anthropocene boundary hypotheses. When the term Anthropocene was originally proposed, a "late" start date was suggested which coincided with the onset of the Industrial Revolution in Great Britain about A.D. 1800 (Crutzen and Stoermer 2000). This start date seemed to be embraced by the geological community at first (e.g., Zalasiewicz et al. 2010, 2011; Steffen et al. 2011; Price et al. 2011). However, a Holocene-Anthropocene boundary corresponding to the beginning of the "Great Acceleration" at A.D. 1945 was subsequently proposed (Ford et al. 2014; Waters et al. 2016). This boundary seemed to gain favor because it was thought to be recognizable worldwide based on the unique presence of radioactive fallout in the geologic record marking the end of WWII. At the other end of the spectrum, Ruddiman (2007, 2013) proposed an "early" Anthropocene start date of 8000–5000 yr B.P. defined by increased atmospheric methane attributed to the advent of agriculture. Some archaeologists have proposed an even earlier start date, suggesting that the Holocene be replaced by, or merged with, the Anthropocene (e.g., Smith and Zeder 2013; Erlandson 2013; Erlandson and Braje 2013). The Anthropocene Working Group of the International Stratigraphic Commission recently voted in favor of the WWII start date (Voosen 2016), but the debate appears to be far from over.

Archaeology is the scientific study of the origin, development, and varieties of human beings and their societies based on material remains (e.g., fossil relics, artifacts, and monuments) found in the geologic record. Archaeology (anthropology) and stratigraphy are inextricably linked (Fig. 2.3), hence it is understandable that archaeologists are arguing for a definition of the Anthropocene Epoch consistent with archaeological stratigraphy. The erosional unconformity separating the undisturbed natural part of the geologic record, from the anthropogenic deposits

Fig. 2.3 Complex stratigraphy at a historical (19th century) archaeological site. Layer of natural soil (*I*) is overlain by anthropogenic sediments (layers *II–IV*). Photo by Don Adzigian

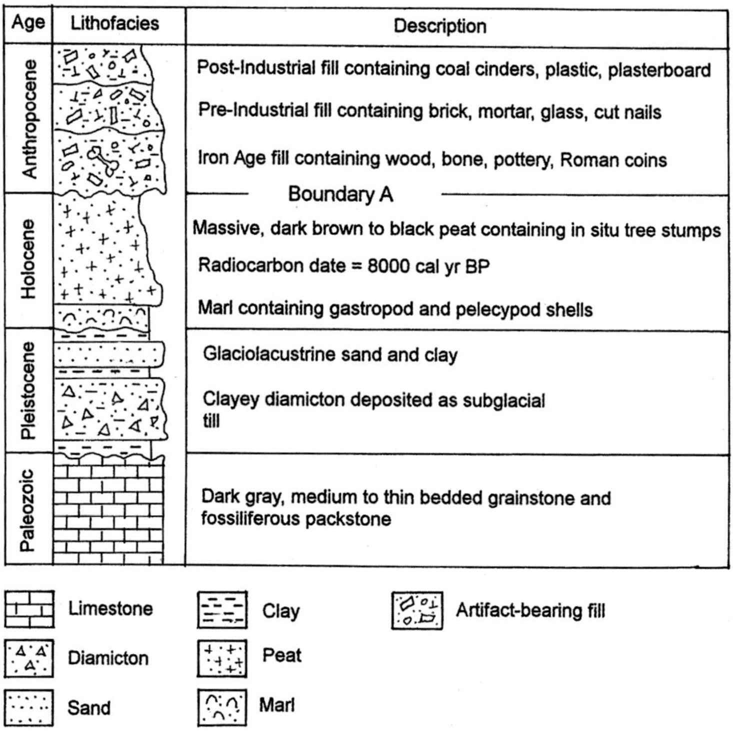

Age	Lithofacies	Description
Anthropocene		Post-Industrial fill containing coal cinders, plastic, plasterboard
		Pre-Industrial fill containing brick, mortar, glass, cut nails
		Iron Age fill containing wood, bone, pottery, Roman coins
		——— Boundary A ———
Holocene		Massive, dark brown to black peat containing in situ tree stumps
		Radiocarbon date = 8000 cal yr BP
		Marl containing gastropod and pelecypod shells
Pleistocene		Glaciolacustrine sand and clay
		Clayey diamicton deposited as subglacial till
Paleozoic		Dark gray, medium to thin bedded grainstone and fossiliferous packstone

Legend: Limestone, Clay, Artifact-bearing fill, Diamicton, Peat, Sand, Marl

Fig. 2.4 Hypothetical stratigraphic column showing the physical basis for recognizing the Anthropocene Epoch as coincident with the historic archaeosphere. Anthropogenic strata lie above a visible unconformity (Boundary A) marking the top of undisturbed natural Holocene sediments

locally comprising the uppermost part of the geologic column, has been referred to by archaeologists as Boundary A (Fig. 2.4). The collective global package of archaeological strata lying above Boundary A comprises the archaeosphere (Edgeworth 2014; Edgeworth et al. 2015). Thus, from an archaeological point of view the archaeosphere is physically equivalent to the Anthropocene Series as a chronostratigraphic unit. Geologically, this is a problem because Boundary A is time-transgressive, and thus fails to meet the requirement for a stratigraphic "golden spike" (Edgeworth et al. 2015).

Regardless of how the Anthropocene is eventually defined, Fig. 2.5 shows that there are multiple archaeological events that coincide with the rise of civilization in Mesopotamia about 3300 B.C., including the development of the first known city-based states and the world's oldest known written historic record (Anthony 2010; Fagan 2011). These events represent multiple, diagnostic archaeological indicators whose stratigraphic ranges converge to define a "time line" at $\sim$6000–5000 cal yr ago in a manner analogous to an assemblage biozone. This "time line" corresponds to the first appearance in the geologic record locally of cities, state formation, social hierarchies, writing, and metallurgy (copper smelting).

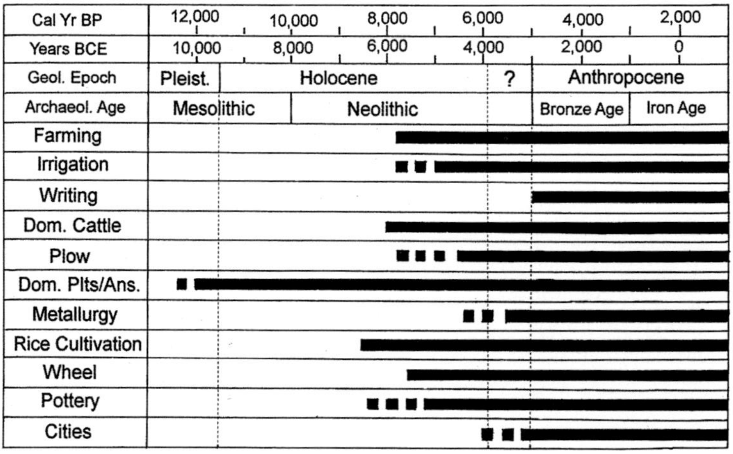

2.4 Stratigraphy and Archaeology

Cal Yr BP	12,000	10,000	8,000	6,000	4,000	2,000
Years BCE	10,000	8,000	6,000	4,000	2,000	0
Geol. Epoch	Pleist.	Holocene		?	Anthropocene	
Archaeol. Age	Mesolithic	Neolithic			Bronze Age	Iron Age
Farming						
Irrigation						
Writing						
Dom. Cattle						
Plow						
Dom. Plts/Ans.						
Metallurgy						
Rice Cultivation						
Wheel						
Pottery						
Cities						

Fig. 2.5 Timeline of archaeological events useful for defining a Holocene-Anthropocene boundary at 6000–5000 cal yr BP coincident with the beginning of civilization in Mesopotamia

Anthropogenic soil erosion and river sedimentation were also occurring in Sumeria by this time (Lowdermilk 1953; Montgomery 2012). As noted above, early cities were able to grow because of the domestication of cattle, and the contemporaneous development of irrigation and the plow about 5500 B.C. The Sumerian city of Uruk is thought to have had a population of about 50,000 by 3000 B.C., and at its peak about 1800 B.C., the Mesopotamian region may have had 15–20 million inhabitants (Lowdermilk 1953). Hence, the "time line" at $\sim$5000 yr BP corresponds to the beginning of the first urban population explosion on Earth. It also marks the beginning of increases in atmospheric CO_2 and CH_4 attributable to human activity (Ruddiman and Thomson 2001). Although Gail and Hoare (2012) argued against the use of anthropogenic soils as a golden spike for the Anthropocene, it seems clear that any artifact-bearing anthropogenic soils and sediments preserved in the geological record eventually will be the best visible physical evidence of the Anthropocene Series.

In addition to proposals to establish an Anthropocene Epoch, it has also been proposed (Howard 2014) that the stratigraphic code be amended to include Anthrostratigraphic units (ASUs) and Technostratigraphic units (TSUs). An ASU is defined as a stratiform or irregularly shaped body of anthropogenic origin in the sedimentary record distinguished and delineated on the basis of lithologic characteristics and/or bounding disconformities. An anthropogenic origin is indicated if the deposit: (1) Lies beneath an anthropogenic landform, (2) Contains artifacts, (3) Shows evidence of artificial mixing or other human disturbance, and/or (4) Was imported from offsite and is allochthonous. The basic unit of anthrostratigraphy is the *anthrostratum*, defined as a mostly stratiform body of artificially mixed earth (rock, sediment, soil, etc.) and artifactual (brick, concrete, etc.) materials. An anthrostratum need only be mappable at a local scale, but anthrostrata of different origins or type may be grouped together as an *anthroformation* if regionally mappable. This terminology is consistent with, and perhaps improves upon, the archaeological "layer" (Gasche and Tunca 1983; Stein 1990), the British Geological Survey's definition of artificial ground (Rosenbaum et al. 2003; Price et al. 2011; Ford et al. 2014), the U.S. National Resource Conservation Service's definition of human-transported material (Soil Survey Staff 2014), and what is generally referred to as archaeological stratigraphy (Harris 1989, 2014)

A TSU was defined as a mostly stratiform body of anthropogenic origin in the sedimentary record identified or characterized on the basis of artifact content (Howard 2014). An artifact was defined as any object of anthropogenic origin including: (1) Demolition wastes (brick, mortar, concrete, wood, glass, etc.), (2) Coal-related wastes (coal, coke, carbonaceous shale, cinders, etc.), (3) Petroleum-related wastes (asphalt, "PetCoke," etc.), (4) Industrial or commercial manufacturing wastes (coal-tar, steel-making slag, etc.), (5) Archaeological objects (stone tools, pottery, coins, etc.), and (6) Miscellaneous rubbish (plastic, cardboard, etc.). The definition included microartifacts, and certain rock materials that can be considered to be artifacts simply because their presence in a site-specific context is due to human activity (Dunnel and Stein 1989). For example, fragments of coal (and associated carbonaceous shale) imported for domestic coal-burning can be considered to be artifacts, whereas detrital limestone and dolostone clasts derived from underlying bedrock are not artifacts.

Zalasiewicz et al. (2014) proposed the term "technostratigraphy" for the use of artifacts (termed technofossils) to date and correlate anthropogenic deposits. In concept, artifacts in TSUs can be used, in a manner analogous to the use of fossils in biostratigraphy, for dating and correlating anthropogenic deposits. TSUs are based on the evolution over time of technology involved in the creation of artificially altered or manufactured objects. That is, anthrostrata in the Quaternary sedimentary record show an overall upsection technological evolution in the artifact assemblages contained in them. For example, "low-tech" objects, pre-dating the Industrial Revolution (Paleolithic stone tools, Neolithic copper, hand-forged wrought-iron nails, etc.), contrast with "high-tech" artifacts of more modern vintage (drywall, galvanized nails, Portland cement, PVC piping, etc.). The date of invention of modern artifacts is known from historic records, thus providing a highly precise dating tool for the anthrostratum. The TSU is useful for chronological analysis of sites too young to be dated by the radiocarbon method. The *technozone* was proposed as the fundamental unit of technostratigraphy, defined on the basis of artifactual (Commercial) ranges (Howard 2014). By analogy with biostratigraphy, an "index artifact" can be used to name a technozone. The TSU supplants the Ethnostratigraphic Unit of previous archaeological classifications (Gasche and Tunca 1983; Stein 1990) which relied on a cultural interpretation of materials making up the unit.

Howard (2014) showed how the Commercial Range (CR) of an artifact could be used as a chronostratigraphic tool, where CR is defined as the time span of availability to society from the beginning to the end of its commercial production (Fig. 2.6). Using the biostratigraphic range of a fossil species as an analogy,

2.4 Stratigraphy and Archaeology

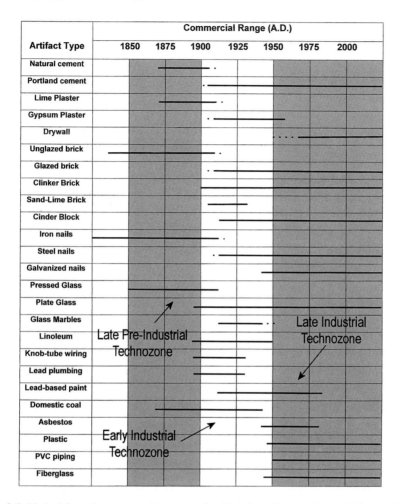

Fig. 2.6 Method for using commercial ranges of artifacts in anthropogenic surficial deposits to define technozones (Howard 2014)

technozones for Detroit, Michigan were defined by first appearances and the overlapping CRs of artifacts in an assemblage. For example, a 19th century anthropogenic deposit from the late pre-Industrial Revolution technozone was identified by an artifactual association of natural hydraulic cement (mortar), unglazed brick, lime-based wall plaster, wrought-iron (square) nails, pressed glass and coal-related wastes. A 20th century early Industrial Revolution technozone was defined by the first appearance of Portland cement (mortar and concrete), gypsum-based wall plaster, glazed brick, steel (round) nails, and glass marbles. An association between linoleum, ceramic tubes (knob-and-tube wiring) and lead plumbing was also diagnostic. The late Industrial technozone was defined by the first appearance of drywall, plastic, PVC piping, fiberglass, etc. Thus, there was a well defined stratigraphic transition from "low-tech" to "high-tech" artifact assemblages in the geologic column of Detroit at ~ A. D. 1900.

References

Anthony DW (2010) The horse, the wheel, and language: how Bronze age riders from the Eurasian steppes shaped the modern world. Princeton University Press, 568 pp
Artzy M, Hillel D (1988) A defense of the theory of progressive soil salinization in ancient southern Mesopotamia. Geoarchaeology 3:235–238
Brevik EC, Hartemink AE (2010) Early soil knowledge and the birth and development of soil science. Catena 83:23–33
Certini G, Scalenghe R (2011) Anthropogenic soils are the golden spikes for the Anthropocene. The Holocene 21:1267–1274
Childe VG (1950) The urban revolution. Town Plann Rev 21:3–17
Crutzen PJ (2002) The geology of mankind. Nature 415:23
Crutzen PJ, Stoermer EF (2000) The "Anthropocene". IGBP Newsl 41:17–18
Davis K (1955) The origin and growth of urbanization in the world. Am J Sociol 60:429–437
Dunnel RC, Stein JK (1989) Theoretical issues in the interpretation of microartifacts. Geoarchaeology 4:31–42
Edgeworth M (2014) The relationship between archaeological stratigraphy and artificial ground and its significance in the Anthropocene. In: Waters CN, Zalasiewicz J, Williams M (eds) A stratigraphical basis for the Anthropocene, vol 395. Geological Society of London Special Publication, pp 91–108
Edgeworth M, deB Richter D, Waters C, Haff P, Neal C, Price NJ (2015) Diachronous beginnings of the Anthopocene: the lower bounding surface of anthropogenic deposits. Anthropocene Rev 21:33–58
Erlandson JM (2013) Shell middens and other anthropogenic soils as global stratigraphic signatures of the Anthropocene. Anthropocene 4:24–32
Erlandson JM, Braje TJ (2013) Archaeology and the Anthropocene. Anthropocene 4:1–7
Fagan BM (2011) World prehistory—a brief introduction, 8th edn. Longman, New York, 408 pp
Ford JR, Price SJ, Cooper AH, Waters CN (2014) An assessment of lithostratigraphy for anthropogenic deposits. Geol Soc London Spec Pub 395:55–89
Gale SJ, Hoare PG (2012) The stratigraphic status of the Anthropocene. Holocene 22:1491–1494
Gasche H, Tunca O (1983) Guide to strratigraphic classification and terminology; Definitions and principles. J Field Archaeol 10:325–335
Harris EC (1989) Principles of archaeological stratigraphy. Academic Press, London 170 pp
Harris EC (2014) Archaeological stratigraphy as a paradigm for the Anthropocene. J Contemp Archaeol 1:105–109
Hooke RL (1994) On the efficacy of humans as geomorphic agents. GSA Today 4(217):224–225
Hooke RL (2000) On the history of humans as geomorphic agents. Geology 28:843–846
Hooke RL, Martin-Duque JF, Pedraza J (2012) Land transformation by humans: a review. Geol Soc Am Today 22:4–10
Howard JL (2014) Proposal to add anthrostratigraphic and technostratigraphic units to the stratigraphic code for classification of anthropogenic Holocene deposits. Holocene 24:1856–1861
Legget RF (1973) Cities and geology. McGraw-Hill, New York 624 pp
Lowdermilk WC (1953) Conquest of the land through 7000 years. Agric Inform Bull 99:43 pp (Soil Conservation Service, U. S. Gov. Printing Office)
Lubowski RN, Vesterby M, Bucholtz S, Baez A, Roberts M (2006) Major uses of land in the United States. Econ Inf Bull EIB-14:54 pp (U. S. Dept. Agric, Economic Res. Serv.)

References

Montgomery DR (2007) Is agriculture eroding civilization's foundation? GSA Today 17:4–9

Montgomery DR (2012) Dirt: the erosion of civilization. University of California Press, Berkeley, CA 285 pp

Oldfield F (2015) Can the magnetic signatures from inorganic fly ash be used to mark the onset of the Anthropocene? Anthropocene Rev 2:3–13

Olson GW (1981) Soils and the environment: a guide to soil surveys and their applications. Chapman and Hall, New York, N.Y. 200 pp

Orndorff RC (2007) Divisions of geologic time—major chronostratigraphic and geochronologic units. U. S. Geological Survey Fact Sheet 2007-3015

Price SJ, Ford JR, Cooper AH, Neal C (2011) Humans as major geological and geomorphological agents in the Anthropocene: the significance of artificial ground in Great Britain. Philos Trans R Soc (Ser A) 369:1056–1084

Rosenbaum MS, McMillan AA, Powell JH, Cooper AH, Culshaw MG, Northmore KJ (2003) Classification of artificial (man-made) ground. Engineering Geology 69: 399–409

Ruddiman WF (2007) The early Anthropocene hypothesis: challenges and responses. Rev Geophy 45(RG4001):37 pp

Ruddiman WF (2013) The Anthropocene. Annu Rev Earth Planet Sci 41:45–68

Ruddiman WF, Thomson JS (2001) The case for human causes of increased atmospheric CH_4 over the last 5000 years. Quat Sci Rev 20:1769–1777

Seto KC, Guneralp B, Hutyra LR (2012) Global forecasts of urban expansion to 2030 and direct impacts on biodiversity and carbon pools. Proc Nat Acad Sci 109:16083–16088

Smith BD, Zeder MA (2013) The onset of the Anthropocene. Anthropocene 4:8–13

Soil Survey Staff (2014) Keys to Soil Taxonomy (12th edition). U.S. Department of Agriculture, Natural Resources Conservation Service, pp 372

Steffen W, Grinevald J, Crutzen P, McNeill J (2011) The Anthropocene: conceptual and historical perspectives. Philos Trans R Soc (Ser A) 369:842–867

Stein JK (1990) Archaeological stratigraphy. In: Lasca NP, Donahue J (eds) Archaeological geology of North America. Boulder, Colorado. Geological Society of America Centennial Special Volume 4, pp 513–523

Syvitski JPM, Kettner A (2011) Sediment flux and the Anthropocene. Philos Trans R Soc (Ser A) 369:957–975

The United Nations (2014) World urbanization prospects: the 2014 revision, Highlights. United Nations, Department of Economic and Social Affairs, Population Division (ST/ESA/SER.A/352)

Voosen P (2016) Anthropocene pinned to post-war period. Science 353:852–853

Waters CN, 23 others (2016) The Anthropocene is functionally and stratigraphically distinct from the Holocene. Science 351:aad2622

Wilkinson BH, McElroy BJ (2007) The impact of humans on continental erosion and sedimentation. Geol Soc Am Bull 119:140–156

Zalasiewicz J, Williams M, Steffen M, Crutzen P (2010) The new world of the Anthropocene. Environ Sci Technol 44:2228–2231

Zalasiewicz J, Williams M, Fortney R, Smith A, Barry TL, Coe AL, Bown PR, Rawson FJ, Gale A, Gibbard P, Gregory FJ, Hounslow MW, Kerr AC, Pearson P, Knox R, Powell J, Waters C, Marshall J, Oates M, Stone P (2011) Stratigraphy of the Anthropocene. Philos Trans R Soc (Ser A) 369:1036–1055

Zalasiewicz J, Williams M, Waters CN, Barnosky AD, Haff P (2014) The technofossil record of humans. Anthropocene Rev 1:34–43

Chapter 3
Anthropogenic Landforms and Soil Parent Materials

Anthropogenic landforms are created either directly by artificial processes (e.g., strip mining), or indirectly by natural processes triggered by human activity (e.g., accelerated soil erosion). They are commonly produced by the building up of the land with artificial fill materials (aggradation), or as a result excavation (degradation). Anthropogenic landforms vary as a function of geocultural setting, and are generally recognized by their deviation from the natural landscape. Anthropogenic landforms can be classified physically into six basic types: (1) flats, (2) benches, (3) terraces, (4) convexomorphic, (5) concavomorphic, and (6) plateau. The NRCS Geomorphic Description System used in the U.S. classifies landforms hierarchically according to landscape, landform, microfeature and anthropogenic feature. Two drawbacks to the system are that use of the term "feature" conflicts with that of archaeological usage, and the term "fill" is used both for a landform and a sedimentary deposit. Alternatively, the British Geological Survey classifies anthropogenic landforms as five distinct types of artificial ground using a hierarchical morphostratigraphic approach. Soil parent material can be used to distinguish two basic types of anthropogenic soils. Metagenetic anthrosoils are formed from human-altered parent material, i.e. pre-existing soils that were modified extensively *in situ* by human action. Neogenetic anthrosoils are formed from human-transported material, i.e. new anthropogenic sediments produced and deposited by an artificial mechanism as a direct result of human activities. Metagenetic anthrosoils are generally characteristic of agricultural settings, and human habitation, ceremonial and grave sites in archaeological settings. Neogenetic anthrosoils are characteristic of ancient burial mounds, and mine-related and urban settings where anthrosediments are often created on a vast scale using earthmoving equipment.

3.1 Introduction

Geologists map bedrock by classifying rock bodies into lithostratigraphic units, i.e. map units defined by the physical (lithologic) characteristics of rocks. The most basic type is the formation, which may be subdivided into a member or bed, or assembled together with other formations to form a group. The procedure is spelled out in the North American Stratigraphic Code of Nomenclature (NACSN 2005).

Mapping surficial deposits requires an approach different from that used to map rocks. It is similar to the way that soil scientists map soils, which is based on slope (landscape position) and soil profile characteristics. Geologists map surficial deposits using a morphostratigraphic (allostratigraphic) approach, i.e., a map unit is delineated as the surficial deposit lying directly beneath a genetically-related, mappable landform. A landform is any recognizable physical form or feature on the Earth's surface having a characteristic shape or slope. In other words, a landform has a specific topographic expression defined by differences in elevation. Several different types of landforms typically make up a landscape, which is defined as the distinct association of landforms that can be seen together in a single view. Anthropogenic soils are usually part of an anthroscape, i.e. a human-modified landscape. Hence, the principles used to classify and map landforms and surficial deposits are critical to an understanding of anthropogenic soils and their parent materials.

3.2 Anthropogenic Landforms

3.2.1 Human Impacts on the Landscape

Anthropogenic landforms are created by human impacts on the landscape that can be categorized generally as either intentional (direct) or unintentional (indirect). Intentional impacts are typically the result of human activities for which there is no natural counterpart, whereas unintentional impacts usually involve a natural process which has been triggered or accelerated by human activity. For example, as discussed in Chap. 2, accelerated soil erosion and river sedimentation was inadvertently caused by ancient agriculture and deforestation. Intentional impacts can be classified as: (1) aggradational or constructional, (2) degradational or destructional, and (3) hydrological. **Aggradational impacts** usually involve the building up or leveling of the land with artificial fill materials transported by mechanized and earthmoving equipment. **Degradational impacts** involve the artificial planation or erosion of the land surface by stripping or excavation of earth materials. Both aggradational and degradational impacts usually have hydrological implications the former causing a relative drop, and the latter an artificial rise, in water table levels. Unintentional impacts are usually the indirect consequences of human activities in the form of: (1) human-induced soil erosion and associated river sedimentation, (2) ground subsidence, (3) slope failures, and (4) earthflows.

3.2 Anthropogenic Landforms

Human impacts on the landscape vary with geocultural setting (Table 3.1). **Urbanization** involves residential and commercial development during which the ground is often leveled, built up or excavated for building construction or landscaping purposes. **Industrial activities** may artificially erode the landscape via construction of underground storage tank facilities, retention ponds, canals and secure landfills, whereas the accumulation of waste materials often results in the creation of aggradational landforms. **Agricultural activities** tend to create degradational landforms, although human-induced soil erosion may cause greatly enhanced river sedimentation. **Transportation** involves the construction of streets, parking structures, highways, railroads, and airports, which typically requires sealing the ground surface with manufactured materials (e.g., concrete). This disrupts soil formation processes and groundwater recharge. **Water management** strategies may involve the construction of aggradational landforms (e.g., levees and dykes) for coastal and stream flood control, or degradational landforms such as drainage ditches and irrigation or flood-control channels. **Surface mining** produces large-scale cut-and-fill landforms of various types, and even underground mining can affect the landscape via surface subsidence or collapse. **Warfare** can result in long-lasting land modifications directly via ramparts, trenches and bomb craters, or indirectly via devegetation and enhanced soil erosion. **Tourism and sports-related activities** such as golfing, skiing and off-road vehicle use can also result in significant landscape modifications locally.

Table 3.1 Human impacts on the landscape. After Szabo et al. (2010)

Human activities	Impacts on landscape	
	Aggradational	Degradational
Urbanization	Surface sealing with manufactured materials, demolition site fill, sanitary landfills, landscaping	Basement excavations, borrow pits, artificial wetlands, slope failures
Industrialization	Raw material storage and waste material piles, secure landfills	Retention ponds, canals, underground storage tanks
Agriculture	Enhanced river sedimentation	Enhanced soil erosion, terrace cultivation, irrigation ditches
Transportation	Highway and railroad embankments, grading fill	Highway and railroad cuts
Water Management and Exploitation	Dams, artificial levees, dykes	Canals, ditches, flood control channels, ground subsidence
Mining	Spoil piles, backfill to restore original topography	Blasting and excavation of bedrock, mountain top removal, ground subsidence
Warfare	Ramparts and other defensive earthworks	Bomb craters, trench excavations
Tourism and sports	Landscaping for hiking and ski trails, golf courses and race tracks	Accelerated soil erosion caused by off-road vehicles and hiking trails, artificial lakes and ponds

3.2.2 Natural Versus Anthropogenic Landforms

Anthropogenic landforms are generally recognized by their deviation from the normal appearance of the natural landscape in any given region. Hence, an understanding of natural landforms is prerequisite for the recognition of human impacts on the landscape. The topographic profiles of natural landforms can generally be divided into four basic components: (1) an upper convex segment, (2) a cliff face or free face, (3) a straight segment, and (4) a concave segment at the base of the slope (Fig. 3.1). The most common slope profile in a humid-temperate region has a distinct convex upper slope and a concave lower slope (Fig. 3.2a). This kind of profile is formed where the rate of weathering is more rapid than erosion, and where chemical weathering predominates over physical weathering. The upper convexity is attributed to soil creep, which predominates over slope wash occurring on the lower concave slope. This slope profile normally develops on unconsolidated parent material regardless of climate, but it is typically dominant in humid-temperate zones where vegetation is continuous. The convexo-concave profile is less affected by parent rock type, and more dependent on the type and rate of slope processes (Ritter et al. 2011).

Slope profiles in arid and semi-arid climates tend to be more angular with a free face predominating over the upper convex segment (Fig. 3.2b). Steep cliffs are often present above a straight, debris covered segment, also known as talus, which normally lies in repose at an angle of $\sim 30°$. A pronounced change in slope occurs at the base of the straight segment, and angles decrease over a short distance to less than $5°$, the typical slope of most desert plains. This slope profile develops where the rate of soil or regolith production is lower than the rate of erosion. Such areas are characterized by shallow rocky soils, and these slope profiles tend to be determined by the character of the bedrock. Rock resistance, jointing, bedding or foliation control the type and extent of loose rock fragments comprising the debris slope. The limited vegetation and predominance of physical weathering and mass wasting limits soil creep to a minimum. Slope wash along the base of the slope is also more important than in humid-temperate climates.

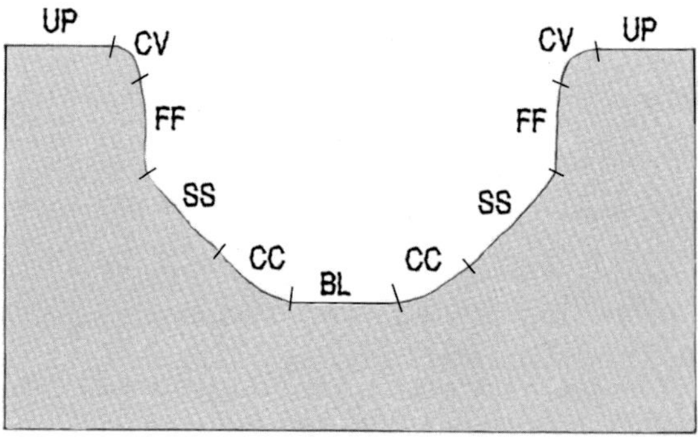

Fig. 3.1 Major components of slope profiles: *UP* upland; *CV* convex-upward; *FF* free-face; *SS* straight segment: *CC* concave-upward; *BL* bottomland

3.2 Anthropogenic Landforms

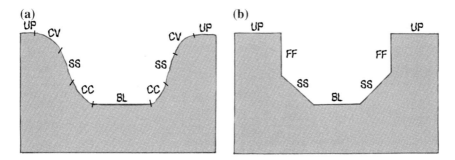

Fig. 3.2 Typical slope profiles in different climatic settings: **a** Humid region; **b** Arid region. *UP* Upland; *CV* convex-upward; *FF* free-face; *SS* straight segment; *CC* concave-upward; *BL* bottomland

Anthropogenic soils tend to be associated with distinctive artificial landforms depending on whether they are located in archaeological, agricultural, mine-related or urban settings. In an archaeological setting, human occupation and ceremonial activities can produce relatively small landforms such as middens and burial mounds. Larger landforms include tells, often built up by thousands of years of human occupation, and geomorphic features associated with the construction of monumental architecture. Agricultural settings are often characterized by hill slope terraces, and landforms created to either drain or irrigate the land. Surface mining for coal and other fossils fuels, precious metals, and sand and gravel can disturb the land surface over very large areas. Mine-related landforms depend on the nature of the mining method. For example, strip-mining for coal often produces a terraced landscape analogous to that seen in hill slope agricultural settings. However, underground mining can also affect the surface topography by way of ground subsidence or collapse. Anthropogenic landforms in urban areas, such as artificial berms, can be classified according to land use as residential, industrial, transportation-related, water management-related, and so forth.

3.2.3 Physical Landform Classification

Anthropogenic landforms can be classified physically into six basic types (Table 3.2; Fig. 3.3): (1) flats, (2) benches, (3) terraces, (4) convexomorphic, (5) concavomorphic, and (6) plateau. Flats are perhaps the most widespread type of anthropogenic landform (Fig. 3.4a), and are most easily identified in hilly or mountainous terrain. They are often created during construction using the cut-and-fill method. Dozers are typically used to grade a site in preparation for construction, in which case the original soil may be more or less stripped off. A soil

Table 3.2 Physical classification of anthropogenic landforms

Landform type	Examples
Flat	Industrial commercial or residential (house pad) building site, parking lot, highway, railway, demolition sites
Bench	Building construction, road emblankment, contour mining
Terrace	Contour mining, hillslope agriculture
Convexomorphic	Berm, burial mound, mine tailings
Concavomorphic	Gravel pit, quarry, open pit mine, artificial wetland
Plateau	Mountaintop removal, tell, landfill

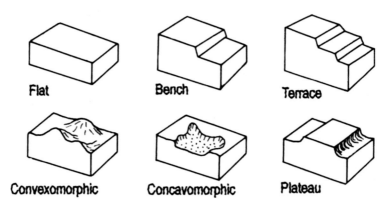

Fig. 3.3 Physical classification of anthropogenic landforms. See Table 3.2 for further explanation

developed subsequently on this artificial cut can be considered to be anthropogenic, but difficult if not impossible to distinguish from a natural soil. Alternatively, a flat can be created following building demolition when dozers are used to backfill and grade a site with mixed earth materials often containing waste building materials. An anthropogenic soil developed on a former demolition site is easily recognized as having developed in an anthropogenic surficial deposit. A bench is sometimes associated with construction or landscaping in urban areas, highway embankments (Fig. 3.4b) or with strip mining. Terraces are associated with hillslope agriculture and strip mined land (Fig. 3.4c). Convexomorphic landforms range from heaps of mine tailings and other anthropogenic wastes (Fig. 3.4d), to berms created for traffic noise abatement, and burial mounds at archaeological sites. Concavomorphic landforms (Fig. 3.4e) are created by digging on scales ranging from a small drainage ditch to a giant open pit mine. Artificial plateaus are characteristic of landfills (Fig. 3.4f), the mountaintop removal method of surface mining, tells, and so forth.

3.3 Genetic Landform Classification Systems

Fig. 3.4 Examples of anthropogenic landforms: **a** Flat formed by building demolition; **b** Bench comprising highway enbankment; **c** Artificial terraces produced by strip-mining for coal; **d** Convexomorphic landform comprised of coal-related waste heaps; **e** Concavomorphic landform created as an artificial wetland; **f** Plateau in the form of an urban landfill

3.3 Genetic Landform Classification Systems

Geomorphologists usually classify natural landforms using a genetic approach, i.e. according to the predominant geomorphic processes responsible for their origin. This approach can be problematic if there is controversy amongst soil scientists or geomorphologists as to how a particular landform was formed, or if subsurface information required for precise identification is unavailable, or unobtainable due to limited property access or other problems. Dubious interpretations can always be queried, or the physical classification outlined in Fig. 3.3 can be used.

Geomorphologists generally interpret landscape development based on the identification of geomorphic surfaces. A **geomorphic surface** is a mappable part of the landscape that has definite geographic boundaries, and is formed by one or more processes during a given span of time. In natural landscapes, both depositional and erosional geomorphic surfaces can be differentiated. **Depositional geomorphic surfaces** are formed as a direct result of some natural process of sedimentation. **Erosional geomorphic surfaces** are formed by denudation (Daniels et al. 1971). Both types of surfaces can be flat or irregular depending on the particular process involved.

3.3.1 NRCS Geomorphic Description System

In the United States, the National Resource Conservation Service (NRCS) uses a hierarchical landform classification system (Table 3.3). The NRCS Geomorphic Description System (GDS) consists of three specific sections (Schoeneberger and Wysocki, 2012): (1) **Physiographic Location** (identifies a location within a specific geographic region, (2) **Geomorphic Description** (identifies a discrete landform, or assemblages of landforms, and dominant geomorphic processes), and (3) **Surface Morphometry** (describes in detail discrete land surface geometry and slope segments).

Table 3.3 Components of the NRCS Geomorphic Description System (Schoenberger and Wysocki 2012; Soil Survey Staff 2013)

Geomorphic descriptor	Description
Physiographic location	
Physiographic division	A large geographic part of a continent characterized by similar climate, geologic structure and geomorphic history as depicted at a small scale (e.g., 1:5,000,000)
Physiographic province	A region within a physiographic division with a distinctive pattern of relief or landforms
Physiographic section	A region within a physiographic province with a distinctive pattern of relief or landforms
State physiographic area	A region of relatively local extent within a physiographic section with a distinctive pattern of relief or landforms depicted at 1:100,000–1:500,000 scale
Local name	A local landform depicted on a 7.5 or 15 min topographic quadrangle
Geomorphic description	
Landscape	A land area comprised of a unique assemblage of landforms
Landform	Any recognizable physical form or feature on the earth's surface with a characteristic shape
Microfeature	Small, local landforms that are too small or delineate at 1:10,000 or smaller scale
Anthropogenic feature	An artificial feature on the earth's surface having a characteristic shape and sedimentary composition formed as a direct result of human action
Surface morphometry	
Slope elevation	The height (in meters) of a point on the earth's surface relative to mean sea level
Slope aspect	The compass bearing (in degrees) that a slope faces when viewed downslope
Slope gradient	The difference in elevation between two points, expressed as a percentage of the distance between those two points, in the direction of overland surface water flow
Slope complexity	The relative uniformity (simple) or irregularity (complex) of the ground surface leading downslope through the point of interest

(continued)

3.3 Genetic Landform Classification Systems

Table 3.3 (continued)

Geomorphic descriptor	Description
Slope shape	The two dimensional topographic profiles measured both perpendicular and parallel to the elevation contours (e.g., linear, convex)
Hillslope-profile position	The discrete slope segments found along a transect running perpendicular to the contour beginning at the upland divide (interfluve), and descending to a lower bounding stream or valley floor (i.e. summit, shoulder, backslope, footslope, toeslope landscape positions)
Geomorphic component	The three dimensional representation of the site in the landscape (i.e. hills, terraces, mountains, flat plains)
Microrelief	Slight variations in the height of the land surface that are too small or intricate to delineate on a topographic or soils map at commonly used map scales (1:24,000, 1:10,000, etc.)
Drainage pattern	The geological configuration of the drainage basin of a trunk stream and its major tributaries (e.g., dendritic, deranged) as viewed on a topographic map

Physiographic Locations are named in the GDS according to a hierarchical scheme (Table 3.4) modified from Fenneman and Johnson's (1946) physiographic classification of the United States. Physiographic Location is further subdivided into five categories: (1) Physiographic Division, (2) Physiographic Province, (3) Physiographic Section, (4) State Physiographic Area, and (5) Local Physiographic or Geographic Name. **Geomorphic Descriptions** are made using a hierarchical scheme (Table 3.5) based on four categories: (1) Landscape, (2) Landform, (3) Microfeature, and (4) Anthropogenic feature (if applicable). Landscapes are broad or unique groups of spatially associated land features, whereas Landforms are discrete, individual features of natural origin which are mappable at common soil survey scales (e.g., 1:12,000). Microfeatures are defined by microrelief, i.e. minor variations in topography involving differences in elevation on a scale of generally less than 10 m. Anthropogenic features (artificial or human-modified) are landforms and microlandforms that are created by human activities, and which are too small to be delineated at typical soil survey map scales. **Surface Morphometry** provides a detailed description of land surface shapes and slope elements based on nine different categories: (1) Elevation, (2) Slope aspect, (3) Slope gradient, (4) Slope complexity

Table 3.4 Example of Physiographic Location description according to the NRCS's Geomorphic Description System (Schoeneberger and Wysocki 2012)

Physiographic location	Example
Division	Appalachian Highlands
Province	Blue Ridge Province
Section	Southern Section
State area (optional)	Great Smoky Mountains
Local name (optional)	Clingmans Dome

Table 3.5 Example of Geomorphic Description according to the NRCS's Geomorphic Description System (Schoeneberger and Wysocki 2012)

Geomorphic description	Example
Landscape	Continental glacier
Landform	End moraine
Microfeature	Patterned ground
Anthropogenic feature (if applicable)	Borrow pit

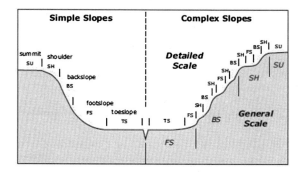

Fig. 3.5 Slope complexity and position according to the Geomorphic Description System used by the USDA-NRCS in the United States (Schoeneberger et al. 2012)

(Fig. 3.5), (5) Slope shape (Fig. 3.6), (6) Hillslope position (Fig. 3.7), (7) Geomorphic component, (8) Microrelief, and (9) Drainage pattern.

Geomorphic descriptions are made using a list of landform terms that are defined (Table 3.6) in the National Soil Survey Handbook Part 629 (Soil Survey Staff 2013). Landscape, landform, and microfeature terms are grouped by geomorphic environment in terms of the predominant geomorphic process responsible for their formation. In addition, the NRCS defines an **anthroscape** as a land area comprised of a unique assemblage of landforms and microfeatures produced by human action. Anthroscapes are typically a complex mosaic of natural landforms, which have been more or less modified by human activity, and new artificial landforms which humans have created by excavation or backfilling. Anthropogenic landforms created as a direct result of anthropogenic deposition or backfilling are classified as **constructional**, whereas those produced by stripping or excavation are **destructional** (Table 3.7). In a geomorphic description of an anthroscape, an anthropogenic landform may be added after, or substituted for, a microfeature term (Table 3.5).

Geomorphic relationships in anthroscapes are generally much more complex than those in natural landscapes, and vary widely as a function of land use history. **Depositional anthropogenic geomorphic surfaces** are formed as a direct result of anthropogenic deposition, e.g., backfilling a demolition site with earthy fill

3.3 Genetic Landform Classification Systems

Fig. 3.6 Terminology used to describe slope shape in the USDA-NRCS system (Schoeneberger et al. 2012)

Down Slope (Vertical)	Across Slope (Horizontal)	Code
concave	concave	CC
concave	convex	CV
concave	linear	CL
convex	concave	VC
convex	convex	VV
convex	linear	VL
linear	concave	LC
linear	convex	LV
linear	linear	LL

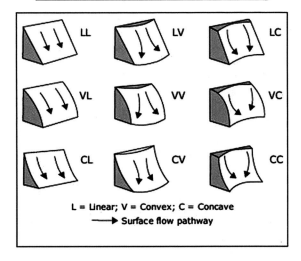

Fig. 3.7 Geomorphic components used to determine hill slope position in the USDA-NRCS system (Schoeneberger et al. 2012)

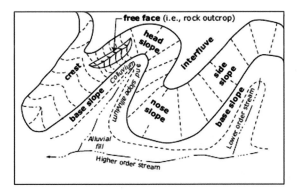

Table 3.6 Types and characteristics of anthropogenic landforms. Modified from U.S. National Soil Survey Handbook Part 629 (Soil Survey Staff 2013)

Anthropogenic landform type	Description
Archaeological	
Tell	A type of archaeological mound created by human occupation and abandonment of a geographical site over many centuries. A classic tell looks like a low, truncated cone with a flat top and sloping sides. The term is mainly used for sites in the Middle East, where it often forms part of the local place name
Burial mound	A small human-made hill comprised of earth which has been heaped up to mark a burial site
Midden	A mound or stratum of refuse (shells, pottery, ash, bones, etc.) normally found on the site of an ancient settlement
Agricultural	
Conservation terrace	An earthen embankment constructed across a slope to prevent soil erosion
Hillslope terrace (ancient)	A raised, generally horizontal strip of earth and/or rock bounded by a down-slope berm or retaining wall, constructed along a contour on a hillslope to prevent accelerated erosion
Leveled land	A field that has been mechanically flattened or smoothed to facilitate management practices such as flood irrigation
Furrow	A shallow channel cut in the soil surface by a plow or disk
Interfurrow	Area of higher soil between two furrows
Ditch	An open and usually unlined trench made to convey water
Drainage ditch	A trench dug to lower the water table in a poorly drainage landscape
Irrigation ditch	A trench dug to provide water to crops
Rice paddy	An anthropogenic, nearly level impoundment that is inundated for long periods typically for wetland rice production. It is applied to areas that have been used in this fashion for a long enough period of time to significantly change the original soil morphology (especially redoximorphic features)

(continued)

3.3 Genetic Landform Classification Systems

Table 3.6 (continued)

Anthropogenic landform type	Description
Dredging and mine-related	
Gravel pit	An excavation to produce gravel usually for concrete or construction
Open pit mine	A relatively large depression resulting from the excavation of material and redistribution of overburden associated with surficial mining operations.
Strip mine	A horizontal strip of rock constructed by blasting along a contour bounded behind by a rock wall; includes level upland surfaces created by mountaintop removal
Reclaimed land	(a) A land area composed of earthy fill material that has been placed and shaped to approximate natural contours, commonly part of land-reclamation efforts after mining operations; (b) A land area, commonly submerged in its native state, that has been protected by artificial structures (e.g. dikes) and drained for agricultural or other purposes (e.g. polder)
Spoil pile	(a) A bank, mound, or other artificial accumulation composed of spoil; e.g., an embankment of earthy material removed from a ditch and deposited alongside it. Compare—dredge spoil bank, (b) A pile of refuse material from an excavation or mining operation; e.g., a pile of dirt removed from, and stacked at the surface of a mine in a conical heap or in layers
Spoil bank	A bank, mound, or other artificial accumulation of rock debris and earthy dump deposits removed from ditches, strip mines, or other excavations
Quarry	Excavation areas, open to the sky, usually for the extraction of stone
High-wall	The rock wall produced by blasting and excavation
Dredged spoil bank	A subaerial mound or ridge comprised of dredged sediment
Artificial collapsed depression	A closed depression in the ground surface formed as a result of surface subsidence or collapse as a result of subsurface mining
Dump	An area of smooth or uneven piles of waste earth material or garbage incapable of supporting plant growth without reclamation
Urban	
Berm	A linear ridge or mound of earthy fill elevated above the surrounding terrain on both sides
Embankment	A raised area of earthy fill used to maintain the grade of a highway or railroad line

(continued)

Table 3.6 (continued)

Anthropogenic landform type	Description
Borrow pit	An excavation from which earth material has been removed usually for construction offsite
Roadbed	The trace or track of a wheeled vehicle route that may or may not be raised slightly above the adjacent land, and composed of earthy fill material (gravel, rock fragments, etc.) or local soil material. Traffic can alter various soil properties primarily by compaction. Abandoned or reclaimed beds may no longer be topographically or visually distinct. However, materials used to construct beds or changes in soil properties may continue to have a significant impact on soil management or plant growth.
Cut	A passage or space from which material has been excavated
Cutbank	A steep slope or earthen wall created by excavation
Sanitary landfill	A land area where municipal solid waste is buried in a manner engineered to minimize environmental degradation. Commonly the waste is compacted and ultimately covered with soil or other earthy material
Artificial levee/dike	An artificial mound of earth constructed along a stream or coastline to prevent flooding
Beveled cut	An artificial slope which has been reduced to prevent mass movements
Floodway	A large-capacity channel constructed to divert floodwaters or excess streamflow from flood-prone urban areas
Scalped area	(a) A modified slope, feature, or land area where much or all of the natural soil has been mechanically removed (e.g. scraped off) due to construction or other management practices. Compare—truncated soil, (b) A forest soil area where the ground vegetation and root mat has been removed to expose mineral soil in preparation for planting or seeding
Railroad bed	The trace or track of a railroad route, commonly constructed slightly above the adjacent land, and composed mostly of earthy materials (gravel, rock fragments, etc.). Abandoned or reclaimed beds may no longer be topographically or visually distinct, but the materials used to construct them may still be a significant portion of the soil zone

3.3 Genetic Landform Classification Systems

Table 3.7 Anthropogenic landforms and microfeatures recognized by the NRCS in the United States. Modified from Schoenberger and Wysocki (2012)

Anthropogenic landforms		Anthropogenic microfeatures	
Constructional	Destructional	Constructional	Destructional
Artificial islands	Beveled cuts	Breakwater	Cutbanks
Artificial levees	Borrow pits	Burial mounds	Ditches
Burial mounds	Canals	Conservation terraces	Furrows
Dumps	Cuts (road, railroad)	Dikes	Hillslope terraces
Dredge-deposit shoals	Cutbanks	Double-bedding mounds	Impact craters
Filled marshland	Dredged channels	Dumps	Skid trails
Earthworks	Earthworks	Embankments	Scalped areas
Fill	Floodways	Fills	
Filled pits	Gravel pits	Hillslope terraces	
Filled enclosures	Leveled land	Interfurrows	
Irrigationally raised land	Log landings	Middens	
Raised land	Open pit mines	Sea walls	
Landfills	Quarries	Rice paddies	
Middens	Rice paddies	Spoil banks	
Mounds	Sand pits	Spoil piles	
Railroad beds	Scalped area		
Reclaimed land	Sewage lagoon		
Rice paddies	Surface mine		
Road beds			
Sanitary landfills			
Spoil banks			
Spoil piles			

materials. These surfaces are often flat, but may be terraced, benched or convexomorphic (Table 3.2). **Erosional anthropogenic geomorphic surfaces** are formed when earth materials are removed as a result of human activities, e.g., excavation of sand and gravel in a borrow pit. They are usually flat or concavomorphic.

Anthropogenic deposits conform to the stratigraphic laws of **superposition** and **cross-cutting relationships**. However, geomorphic surfaces may not conform to the geomorphic **Rule of Ascendency** (landforms at a higher elevation are older than those at a lower elevation) because of human activities and differences in land use history. Assuming that the present land surface was formed at the same time as the fill lying directly beneath it, the relative ages of depositional anthropogenic geomorphic surfaces can be determined using soil profile development. The numerical ages of such surfaces can also be estimated using artifact assemblages as discussed in Chap. 2, or based on historical records, often with much greater precision than is possible with natural depositional geomorphic surfaces. This approach is essential

Fig. 3.8 Complex anthropogenic stratigraphy at an archaeological site in the Corktown district of Detroit, Michigan. Note that the outlines of the surfaces of various archaeological features (*vertical stratigraphic discontinuities*) were scratched onto the outcrop for photographic purposes by archaeologists during the excavation. Photo by Don Adzigian

where anthropogenic sedimentary deposits are too young to be dated using the radiocarbon method.

One drawback of the NRCS system, is that usage of the term "feature" in the GDS conflicts with that of archaeologists who use the term "feature" to describe a *stratigraphic* relationship in an excavation involving **context**, i.e. a layer (anthrostratum) representing a distinct stratigraphic event. An **archaeological feature** has vertical dimensions with respect to site stratigraphy (Fig. 3.8), and typically represents some sort of *in situ* human activity or man-made structure (pits, ditches, walls, etc.). Horizontal characteristics of an archaeological stratigraphic sequence (e.g., the floor of a former dwelling) are not described as features, which also differ from artifacts in that they cannot be separated from their location without changing their form (Schiffer 1987). Another drawback is that the term "fill" is used for both a landform, and the underlying anthropogenic sedimentary deposit.

3.3.2 British Geological Survey Classification

The British Geological Survey uses a four-tiered genetic system to classify anthropogenic landforms as part of what they call "artificial ground." In theory, mapping artificial ground requires recognition of a diagnostic anthropogenic landform, and the physical, sedimentological and lithological characteristics of associated surface deposits. Artificial ground is defined as an area where material is known to have been deposited by humans on the pre-existing land surface, or as an

3.3 Genetic Landform Classification Systems

Table 3.8 Genetic classification scheme for geologic mapping of anthropogenic landforms according to the British geological survey (Rosenbaum et al. 2003)

Hierarchical level	Example	Stratigraphic analogy
Level 1	Artificial ground	Group
Level 2 (class)	Made ground	Formation
Level 3 (type)	Embankment	Member
Level 4 (deposit)	Undifferentiated fill	Lithosome

area where the pre-existing land surface is known to have been excavated. The hierarchical scheme links principal genetic geologic map unit category (Level 1) with basic class (Level 2) and more detailed mapping categories (Level 3). They consider the three levels to be analogous to the stratigraphic units of group, formation and member (Table 3.8), respectively (Rosenbaum et al. 2003). Level 4 relates to the lithology of the deposit, and is used in parallel with the morphostratigraphic divisions of Levels 1 through 3. A three-fold enhanced scheme (Price et al. 2011) is also used to further differentiate classes of artificial ground into type and unit categories.

Five basic classes of artificial ground are recognized (Table 3.9). Anthropogenic deposition of natural earthy and artificial materials typically results in a constructional or aggradational landform. An area is mapped as **made ground** if the fill was deposited on a pre-existing natural land surface. This includes highway and railroad embankments, piles of mine spoil or dredging and artificial levees. **Infilled ground** is associated with similar landforms, but is distinguished by material having been deposited on a preexisting land surface which had been excavated. **Worked ground** refers to places where the land surface had been excavated, but not backfilled, resulting in the creation of an artificial erosional or degradational landform. **Disturbed ground** encompasses areas of artificial ground subsidence caused by underground mining, or by the extraction of groundwater or oil and gas.

Table 3.9 Classes of artificial ground according to the British geological survey (Rosenbaum et al. 2003)

Classes of artificial ground	Description
Made ground	Areas where the ground is known to have been artificially deposited on the former, natural ground surface
Worked ground	Areas where the ground is known to have been artificially cut away or excavated
Infilled ground	Areas where the ground has been artificially cut away (excavated) and subsequently partially or wholly backfilled
Disturbed ground	Areas of surface or near-surface mineral workings with complexly associated excavation and land subsidence features
Landscaped ground	Areas where the original ground surface has been extensively modified, but delineation of made ground, worked ground or disturbed ground is impractical

Landscaped ground refers to land areas that have been modified for aesthetic purposes, and are too small to be delineated at normal geologic map scales.

Although the British classification system works well for land areas deliberately modified by human activities, it stops short of classifying artificial landforms created as a result of unintentional human actions. Hence, Peloggia et al. (2014) proposed a "technogenic" landform classification which may be applicable to areas affected by human-induced erosion (e.g., landslides) or sedimentation (e.g., accelerated soil erosion and fluvial sedimentation). The problem with the Peloggia approach is that, in the absence of artifacts, it may be difficult or impossible to distinguish anthropogenic from natural sediments in the field.

3.4 Anthropogenic Soil Parent Materials

3.4.1 Pedological Description of Parent Materials

Two basic types of anthrosoils can be distinguished based on type of parent material. **Metagenetic anthrosoils** are formed *in situ* from pre-existing soils which are modified more or less extensively by human action. **Neogenetic anthrosoils** are formed from new anthropogenic sediments produced and deposited by an artificial mechanism as a direct result of human activities. Metagenetic anthrosoils are typically found in agricultural settings, but they are also characteristic of human habitation and grave sites in archaeological settings. Neogenetic anthrosoils are characteristic of mine-related and urban settings where anthrosediments are often created on a vast scale using earthmoving equipment. Mine-related, neogenetic anthrosoils are generally comprised exclusively of natural rock and/or soil materials, whereas urban neogenetic anthrosoils often contain abundant artificial components, such as waste building materials produced by building demolition.

In natural soils formed from alluvial, glacial, eolian and other types of sediments, there is often a C horizon (partially weathered parent material) with properties that grow progressively less pedogenic, and more geogenic, with increasing depth. The upper part of the C horizon often shows some weak evidence of leaching and oxidation, then grades downward into an unleached-oxidized zone (Cox horizon of Birkeland 1974), and ultimately into an unleached-unoxidized zone (R horizon, unweathered parent material). It has been suggested previously (Schaetzl and Anderson 2005) that the leached-oxidized zone, showing some evidence of pedogenesis, should be designated as "C horizon," and the unleached-oxidized part be called "D horizon" (Fig. 3.9). Although this terminology has not yet been adopted by the NRCS or other national soil surveys, it seems to be a potentially useful designation. Weathering and leaching will eventually destroy any

3.4 Anthropogenic Soil Parent Materials

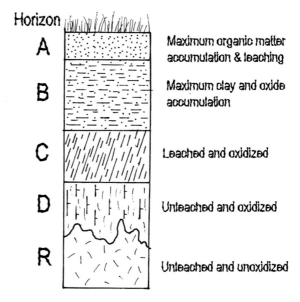

Fig. 3.9 Characteristics of soil profile including a "D horizon."

stratification or other sedimentary fabric originally present in the parent sediment. Thus, the presence or absence of lamination or bedding can perhaps be useful for distinguishing between D and C horizons, respectively. However, care must be taken when applying this criterion because bioturbated sediments, and those deposited by mass flow, typically lack primary stratification.

Neogenetic anthrosoils in mine-related and urban settings can form in anthrosediments that range widely in age. The age of recently deposited anthrosediments may be measured on a scale of weeks, months or a few years, whereas others may have been deposited many decades and perhaps even centuries ago. As long as the deposits remain unsealed and unburied at the surface, pedogenic processes will act to develop a new anthropogenic soil profile. Thus, neogenetic soils grade from the most weakly developed type of profile defined simply based on the presence of growing plants, and thus essentially indistinguishable from anthrosediment, to soils many decades and perhaps even a few centuries old characterized by a relatively well developed A horizon, and possibly even a B horizon in various stages of development. The part of the C horizon in an anthrosoil that shows some evidence of pedogenesis can be regarded as part of the soil profile and described using standard soil survey methods (e.g., Schoeneberger et al. 2012). However, the D horizon, or geogenic part of the C horizon, and the R horizon are best described using a sedimentological approach. Thus, both pedological and sedimentological methods may be required to fully describe an anthropogenic soil profile.

3.4.2 Sedimentological Description of Parent Materials

Hypothetically, anthrosediments can be named and classified with the same methods that sedimentologists use to make lithologic descriptions of rocks and sediments. Thus, physical properties such as color, texture, bedding, mineralogical composition, clast types and abundance, organic matter content, etc. are all potentially useful for describing anthrosediments both in the field and in the laboratory. However, some modifications of existing classifications are needed to accommodate the effects of human action, and the fact that artifacts (artificial objects) are often present, possibly in great abundance. Artifacts are diagnostic of an anthropogenic origin, and they are particularly useful for classifying anthropogenic sediments because they are typically related to land use history.

Clastic sedimentary rocks are classified and named according to texture and mineralogical composition (Folk 1954, 1974). They are distinguished initially by the relative proportions of gravel, sand and mud (silt and clay), using the Udden-Wentworth grade scale (i.e. as rudite, arenite, wacke, and mudrock). Sedimentary rocks are then differentiated further based on mineralogical compositional (e.g., quartz arenite, feldspathic arenite, lithic arenite, etc.). Similarly, natural sediments can be distinguished initially by texture (e.g., gravel, sandy gravel, muddy gravel, sand, etc.), and classified further using mineralogical descriptors (e.g., quartzose, quartzofeldspathic, etc.). Sedimentological classifications of clastic rocks and sediments place considerable emphasis on the presence of any primary mud or clay matrix because this distinguishes mass flow deposits from matrix-free waterlaid sediments.

Anthropogenic sediments are often formed by artificial mixing, hence they frequently contain matrix as a primary feature. They are distinguished as unnatural mixtures of mud, gravel and sand, and by a comingling of natural and artificial types of coarse fragments. Such mixed deposits are best described lithologically by the nongenetic term **diamicton**, a very poorly sorted mixture of gravel, sand, silt and clay in highly variable proportions. Thus, the proposed textural classification for natural sediments shown in Fig. 3.10 was developed by modifying previous classifications (Folk 1954; Raymond 2002) in order to accommodate the use of the term diamicton. Previous studies of urban and archaeological sites show that artifacts are commonly present in both the gravel (macroartifacts) and sand (microartifacts) fractions (Howard et al. 2015; Howard and Orlicki 2015, 2016) of anthropogenic soils. Hence, the textural classification for anthrosediments proposed here (Fig. 3.10) is a modified version of that for natural sediments in which the gravel fraction includes both natural rock clasts (coarse fragments) and artifacts. Anthrosediment names were created by adding the prefix anthro- to indicate an anthropogenic origin. The term "anthromicton" is introduced to account for diamicton of anthropogenic origin. Thus, six distinct types of anthrosediment are recognized: anthrogravel, gravelly anthrosand, anthrosand, anthromicton, muddy anthrosand, and anthromud.

3.4 Anthropogenic Soil Parent Materials

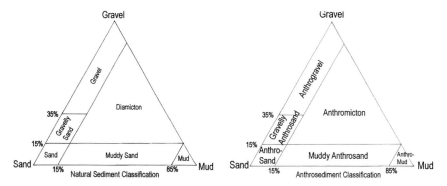

Fig. 3.10 Textural classification of natural and anthropogenic sediments

Anthrosediments containing $\geq 15\%$ gravel may be classified further on the basis of gravel-sized artifact content as shown in Table 3.10. This corresponds to the system used by the NRCS in U.S. Soil Taxonomy to describe anthropogenic soils (Schoenberger et al. 2012), except that a "slightly artifactual" category was added because the presence of any artifacts is indicative of an anthropogenic origin. For example, an anthromicton containing 30% artifacts may be described as an artifactual anthromicton. The proposed compositional classification for anthrosediments containing <15% gravel is shown in Fig. 3.11 and Table 3.11. This classification is modified from some commonly used QFL diagrams used for sandstone classification (Dott 1964; Folk 1974) by combining Q (quartz) and F (feldspar), and emphasizing the presence of sand-sized microartifacts. Anthrosediments containing <15% mud are classified as anthrosands, whereas those with 15–85% mud are muddy anthrosands (analogous to wacke or loamy soil), and those with >85% mud are anthromuds. An anthrosand containing <50% sand-sized lithic grains and 25% microartifacts is an artifactual quartzofeldspathic anthrosand.

Table 3.10 Classification of anthrosediments containing $\geq 15\%$ gravel on the basis of artifact content

Artifact content (% by volume)	Description
0–15	Slightly artifactual
15–35	Artifactual
35–60	Very artifactual
60–90	Extremely artifactual

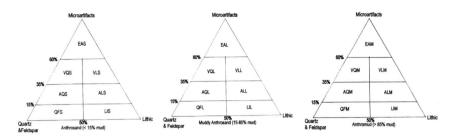

Fig. 3.11 Compositional classification of anthropogenic sediments. See Table 3.11 for further explanation

Table 3.11 Classification of anthrosediments based on composition

Type	Description	Type	Description
Anthrosand			
EAS	Extremely artifactual anthrosand		
VQS	Very artifactual quartzofeld. anthrosand	VLS	Very artifactual lithic anthrosand
AQS	Artifactual quartzofeld. anthrosand	ALS	Artifactual lithic anthrosand
QFS	Quartzofeldspathic anthrosand	LIS	Lithic anthrosand
Muddy anthrosand			
EAL	Extremely artifactual muddy anthrosand		
VQL	Very artifactual quartzofeld. muddy anthrosand	VLL	Very artifactual lithic muddy anthrosand
AQL	Artifactual quartzofeld. muddy anthrosand	ALL	Artifactual lithic muddy anthrosand
QFL	Quartzofeldspathic muddy anthrosand	LIL	Lithic muddy anthrosand
Anthromud			
EAM	Extremely artifactual anthromud		
VQM	Very artifactual quartzofeld. anthromud	VLM	Very artifactual lithic anthromud
AQM	Artifactual quartzofeld. anthromud	ALM	Artifactual lithic anthromud
QFM	Quartzofeldspathic anthromud	LIM	Lithic anthromud

3.5 U.S. Soil Taxonomy

3.5.1 Types of Anthropogenic Parent Materials

The system of Soil Taxonomy used by the NRCS in the United States currently recognizes two basic types of anthropogenic soil parent materials (Soil Survey Staff 2014): (1) **Human-altered material** (HAM), and (2) **Human-transported material** (HTM). Hence, an anthropogenic soil is defined as a soil that has formed either in HAM or HTM. HAM has been artificially mixed or disturbed by humans to a depth of ≥ 50 cm, but shows no evidence of having been transported from outside the pedon. HAM may be composed of either organic or mineral matter. It may contain artifacts such as shells or bones which were used as agricultural amendments. HAM is characteristic of soils that have been mixed as a result of gardening or other horticultural activities, and agricultural soils that have been deeply plowed or ripped to disrupt a root-limiting layer. It is also characteristic of soils that have been excavated and backfilled, as in the case with gravesites in cemeteries, or excessively compacted in order to promote artificial ponding, as in a rice paddy setting. HAM lacks diagnostic subsurface horizons other than the densic horizon commonly found in rice paddy soils.

HTM is a layer of organic and/or mineral matter ≥ 7.5 cm thick that shows evidence of purposeful transport by humans, and that it did not originate from the same pedon which it overlies. The HTM layer rests unconformably on *in situ* material (e.g., a buried A horizon), and frequently contains artifacts (e.g., concrete) not used as agricultural amendments. HTM also may contain mechanically detached pieces of diagnostic horizons, or saprolite, that are allochthonous with respect to the underlying materials. HTM is characteristic of constructional anthropogenic landforms and microfeatures, but may also be found within the confines of a destructional anthropogenic landform or microfeatures (e.g., a borrow pit). Examples of HTM include river dredgings deposited on a pre-existing land surface, layers of fill deposited at a former demolition site, and earthy fill comprising an artificial levee, burial mound or tell. HTM at urban, industrial and mine-related sites may be contaminated with hazardous wastes.

3.5.2 Artifacts

An **artifact** is defined by soil scientists and archaeologists as any artificial object present in soil that is >2 mm in size (Dunnell and Stein 1989; IUSS Working Group 2015; Soil Survey Staff 2014). They are referred to as discrete artifacts if they are persistent, cohesive, and not compacted into a densic horizon that impedes root growth or water movement. They are basically treated as a special variety of coarse fragment, although discrete artifacts which are non-persistent and/or non-cohesive are not considered as coarse fragments. Artifacts are important because they may impede root growth or water movement, and they may contribute significantly to the trace element and total organic carbon content of the soil. Artifacts are a common component of anthropogenic soils, and varying according to geocultural setting.

Artifacts are described in terms of kind, quantity, roundness, shape, cohesion, penetrability, persistence and safety (Tables 3.12 and 3.13). **Kind** refers to the type of object or material being described. **Quantity** is estimated based on volume percentage, or by sieving to obtain a weight percentage. **Roundness** is an expression of the sharpness of the edges or corners of the artifact. **Shape** is a description of overall shape in terms of the relative lengths of the long, intermediate and short particle axes. **Cohesion** is the relative ability of the artifact to resist

Table 3.12 Common kinds of artifacts recognized by U.S. soil taxonomy (Soil Survey Staff 2014)

Type of artifact	Type of Artifact
Bitumen (asphalt, coal)	Fly ash
Boiler slag	Glass
Bottom ash	Metal
Brick	Paper
Cardboard	Plasterboard
Carpet	Plastic
Cloth	Potsherd
Coal combustion products	Rubber (tires, etc.)
Concrete (fragments)	Treated wood
Debitage (stone tool flakes)	Untreated wood

Table 3.13 Terminology used to describe artifacts in anthropogenic soils, according to soil taxonomy (Schoenberger et al. 2012)

Artifact roundness	
Roundness class	**Definition**
Very angular	Strongly developed faces with very sharp, broken edges
Angular	Strongly developed faces with sharp edges
Subangular	Detectable flat faces with slightly rounded corners
Subrounded	Detectable flat faces with well-rounded corners
Rounded	Flat faces absent or nearly absent with all corners rounded
Well rounded	Flat faces absent with all corners rounded
Artifact shape	
Shape class	**Definition**
Elongated	One dimension (length, width or height) is three times longer than either of the others
Equidimensional	Length, width and height are approximately the same
Flat	One dimension is <1/3 that of either of the others, and one dimension isless than three times that of the intermediate axis length
Irregular	Branching or convoluted form
Artifact cohesion	
Cohesion class	**Definition**
Cohesive	Cannot be readily broken into <2 mm pieces
Non-cohesive	Easily broken into <2 mm pieces by hand or simple crushing
Penetrability	
Penetrability class	**Definition**
Non-penetrable	Roots cannot penetrate through or between artifacts
Penetrable	Roots can penetrate through or between artifacts

(continued)

3.5 U.S. Soil Taxonomy

Table 3.13 (continued)

Persistence	
Persistence class	**Definition**
Non-persistent	Susceptible to relatively rapid weathering or decay (expected loss in <10 years)
Persistent	Expected to remain intact in soil for >10 years

Safety	
Safety class	**Definition**
Innocuous artifacts	Harmless to living beings
Noxious artifacts	Potentially harmful or destructive to living beings

mechanical deformation. Non-cohesive artifacts are easily broken down into sand-sized or finer material. **Penetrability** is the relative ease with which roots can penetrate the artifact. **Persistence** is the relative ability of the artifact to withstand weathering and decay over time. **Safety** is the relative toxicity of the artifact to living organisms.

3.6 World Reference Base for Soil Resources

In the early 1980s, the Food and Agriculture Organization of the United Nations (FAO) decided that a framework was needed through which existing soil classification systems could be correlated and harmonized. Meetings were held in Bulgaria in 1980 and 1981 between soil scientists from various programs of the United Nations and the International Society of Soil Science. It was decided to launch a program to develop an International Reference Base for Soil Classification (IRB) with the aim to reach agreement on the major soil groupings to be recognized at a global scale, as well as criteria to define and separate them. In 1992, the IRB was renamed the World Reference Base for Soil Resources (WRB). The first edition of the WRB was published in 1998, and adopted by the International Society of Soil Science as its officially recognized terminology to recognize and classify soils. The WRB is similar to Soil Taxonomy in that it is based mainly on soil morphology rather than genetic concepts of pedogenesis. A major difference is that soil climate is not used as a differentiating characteristic in the WRB. The WRB is meant for correlation of national and local systems, therefore diagnostic criteria tend to closely match those of existing national soil classification systems.

The current edition of the WRB (IUSS Working Group 2015) distinguishes two types of anthropogenic soils, Anthrosols and Technosols. Anthrosols are basically equivalent to HAM-type (metagenetic), and Technosols to HTM-type (neogenetic), soils as defined above. Anthrosols are soils which have been modified profoundly because of irrigation, cultivation and addition of organic materials, or artificial ponding. They are found wherever people have practiced agriculture for many centuries or thousands of years. Some Anthrosols have developed a distinctive human-induced mineral surface horizon that builds up gradually through continuous application of irrigation water with substantial amounts of sediment, and which may include fertilizers, soluble salts, organic matter, etc. Such soils are associated with ancient civilizations found in Mesopotamia and India, and in oases in desert regions. Other Anthrosols have enormously thick A horizons (100 cm or more in thickness) which developed where humans had fertilized the soil with household wastes and manure for many centuries. Such "plaggen" soils are widespread in England, the Netherlands, and other parts of western Europe, and they are found in association with ancient Amazonian, Mayan and Aztec civilizations. Soils affected by artificial ponding, as in rice paddies, occupy vast areas in China, Vietnam, Thailand and Indonesia.

Technosols are soils whose properties and pedogenesis are dominated by their "technological" origin. They contain a significant amount of artifacts reflecting waste materials generated since the beginning of the Industrial Revolution, or perhaps much earlier, e.g., to the time of ancient Rome. Technosols may be sealed by "technic hard material" (manufactured layer), and they are typical of urban and mine-related geocultural settings. Technosols may contain toxic materials, especially in industrial settings. They include soils containing wastes from landfills, sludge, cinders, mine spoils, coal ash, etc.

3.7 British Geological Survey System

Level 4 in the British Geological Survey artificial ground classification relates to the lithology of the deposit, and is used in parallel with the morphostratigraphic divisions of Levels 1 through 3 (Table 3.8). Where there is uncertainty regarding the nature of the fill underlying an anthropogenic landform, it is mapped as undifferentiated fill. Where it is known, it is classified according to the predominant type of waste materials present, or at least whether it is hazardous, non-hazardous, or inert (Rosenbaum et al. 2003; Ford et al. 2014).

References

Birkeland PW (1974) Pedology, weathering and geomorphological research. Oxford University Press, New York

Daniels RB, Gamble EE, Cady JG (1971) The relation between geomorphology and soil morphology and genesis. Adv Agron 23:51–88

Dott RH (1964) Wacke, greywacke and matrix—what approach to immature sandstone classification? J Sediment Petrol 34:625–632

Dunnel RC, Stein JK (1989) Theoretical issues in the interpretation of microartifacts. Geoarchaeology 4:31–42

Fenneman NM, and Johnson DW (1946) Physiographic Divisions of the United States, U.S. Geological Survey, Washington, D.C.

References

Folk RL (1954) The distinction between grain size and mineralogical composition in sedimentary rock nomenclature. J Geol 62:344–359

Folk RL (1974) Petrology of sedimentary rocks. Hemphill, Austin, TX, 182 pp

Ford JR, Price SJ, Cooper AH, Waters CN (2014) An assessment of lithostratigraphy for anthropogenic deposits. In: Waters CW, Zalasiewicz JA, Williams M, Ellis M, Snelling A (eds) A stratigraphical basis for the anthropocene, vol 395, Geological Society of London Special Publication, pp 55–89

Howard JL, Orlicki KM (2015) Effects of anthropogenic particles on the chemical and geophysical properties of urban soils, Detroit, Michigan. Soil Sci 180:154–166

Howard JL, Orlicki KM (2016) Composition, micromorphology and distribution of microartifacts in anthropogenic soils, Detroit, Michigan USA. Catena 138:38–51

Howard JL, Ryzewski K, Dubay BR, Killion TK (2015) Artifact preservation and post-depositional site-formation processes in an urban setting: a geoarchaeological study of a 19th century neighborhood in Detroit, Michigan. J Archaeol Sci 53:178–189

IUSS Working Group (2015) World reference base for soil resources 2014 (update 2015), international soil classification system for naming soils and creating legends for soil maps. World Soil Resources Reports No. 106. FAO, Rome, Italy

NACSN (North American Commission on Stratigraphic Nomenclature) (2005) The North American stratigraphic code: American association petroleum geologists bulletin vol 89, pp 1547–1591

Peloggia AUG, Silva ECN, Nunes JOR (2014) Technogenic landforms: conceptual framework and application to geomorphic mapping of artificial ground and landscape as transformed by human geological action. Quat Environ Geosci 5:67–81

Price SJ, Ford JR, Cooper AH, Neal C (2011) Humans as major geological and geomorphological agents in the anthropocene: the significance of artificial ground in Great Britain. Philos Trans R Soc (Ser A) 369:1056–1084

Raymond LA (2002) Petrology: the study of igneous, sedimentary and metamorphic rocks. McGraw-Hill, New York, 720 pp

Ritter DF, Kochel RC, Miller JR (2011) Process geomorphology. Waveland Press, Long Grove, IL 652 pp

Rosenbaum MS, McMillan AA, Powell JH, Cooper AH, Culshaw MG, Northmore KJ (2003) Classification of artificial (man-made) ground. Eng Geol 69:399–409

Schaetzl RJ, Anderson S (2005) Soils: genesis and geomorphology. Cambridge University Press, 817 pp

Schiffer MB (1987) Formation processes of the archaeological record. University of New Mexico Press, Albuquerque, NM 428 pp

Schoeneberger PJ, Wysocki DA (2012) Geomorphic description system, version 4.2. Natural Resources Conservation Service, National Soil Survey Center, Lincoln, NE

Schoeneberger PJ, Wysocki DA, Benham EC, Soil Survey Staff (2012) Field book for describing and sampling soils, version 3.0. Natural Resources Conservation Service, National Soil Survey Center, Lincoln, NE

Soil Survey Staff (2014) Keys to soil taxonomy, 12th edn. U.S. Department of Agriculture, Natural Resources Conservation Service, 372 pp

Soil Survey Staff (2013) National soil survey handbook Part 629: Glossary of landform and geologic terms. U.S. Department of Agriculture, Natural Resources Conservation Service

Szabo J (2010) Anthropogenic geomorphology: subject and system. In: Szabo J, David L, Loczy D (eds) Anthropogenic geomorphology. Springer, Dordrecht, pp 3–10

Chapter 4
Human Impacts on Soils

Pedogenesis (soil formation) generally results from weathering and horizonation processes. Weathering includes processes that change the physical and chemical characteristics of soil particles, aggregates and artifacts. Horizonation results from additions, losses, translocations, and transformations of solid and chemical soil constituents. Both weathering and horizonation are affected by the five soil forming factors: climate, landscape position, organisms, parent material and time. Anthropogenic soils are formed by anthropedogenic processes defined as artificial conditions that differ from, or significantly modify, the natural processes or factors of soil formation. Human activities often accelerate or disrupt pedogenesis, and some anthropedogenic processes, such as sealing the ground surface with pavement, have no counterpart in the natural world. The combined effects of human activities on the five soil forming factors are so profound that some pedologists regard human activities as a sixth soil forming factor. Although human activities sometimes cause accelerated soil formation, many anthropogenic soils have developed in modern landscapes perhaps no more than a few decades old, and therefore show relatively limited pedogenic development. However, well developed anthropogenic soils are found on ancient landforms such as burial mounds which date back for thousands of years.

4.1 Introduction

Under natural conditions, soil formation results from two basic processes: (1) Weathering, and (2) Horizonation. **Weathering** involves physical processes (e.g., ice wedging) that mechanically disintegrate rocks into progressively smaller sized particles, and chemical processes (e.g., hydrolysis) which decompose and alter rock-forming minerals into new, usually clay-sized, minerals via protonation reactions. Biological processes such as plant root growth can also contribute to physical weathering, and excretion of organic acids from plant roots can promote chemical weathering. **Horizonation** is caused by a wide variety of soil-forming processes (Table 4.1) which can be grouped together into four basic categories (Simonson 1959; Fanning and Fanning 1989): (1) additions, (2) losses, (3) translocations, and (4) transformations. **Additions** (e.g., cumulization) involve the incorporation of mineral, organic or chemical mass into the soil from material

Table 4.1 Some soil-forming processes that operate in natural soils (modified from Buol et al. 2011)

Soil forming process	Description
Additions to the soil	
Enrichment	General term for the addition of material to a soil
Cumulization	Additions of mineral particles to the soil surface by eolian, fluvial or other processes
Melanization	Darkening of the surface horizon by added organic matter
Littering	The accumulation of organic litter on the ground surface
Losses from the soil	
Erosion	Removal of solid material from the top of the surface horizon
Leaching	Removal of soluble ions from part or all of the soil by percolating water under the influence of gravity
Translocations within the soil	
Eluviation	Movement of solid particles downward and away from the upper part of the soil by percolating water under the influence of gravity
Illuviation	Movement of solid particles into a lower part of the soil by percolating water under the influence of gravity
Lessivage	Migration of clay from A to B horizons
Calcification	Accumulation of pedogenic calcium carbonate
Decalcification	Eluviation and perhaps complete removal of pedogenic carbonate from the soil
Salinization	Accumulation of pedogenic chloride and sulfate salts at a rate faster than they can be leached
Desalinization	Removal of soluble salts from part or all of the soil by leaching
Pedoturbation	Mixing of the soil by biological or physical processes
Podzolization	Migration of insoluble iron and aluminum oxides from A to B horizon
Gleization	Reduction and migration of iron under reducing conditions
Transformations within the soil	
Decomposition	The chemical breakdown of mineral and organic materials
Synthesis	The formation of new particles of mineral or organic materials
Humification	The alteration of organic compounds into humus
Mineralization	The release of oxides by decomposition of organic matter
Loosening	Increase in the volume of soil voids by plant and animal activity
Hardening	Decrease in volume of voids by collapse, compaction and filling of open space
Rubification	Reddening of the soil by iron oxides

4.1 Introduction

added to the ground surface by sedimentary or organic processes. **Losses** typically occur by surface erosion of solid particles, or by the leaching of ions or particulate matter from the soil by percolating water moving downward under the influence of gravity. **Translocations** (e.g., lessivage, eluviation, illuviation) refer to the movement of solid particles from one part to another part of the soil, e.g., the downward movement of clay or organic matter from the upper to the lower part of the soil. **Transformations** include chemical weathering and the alteration of one mineral to another, and the biodecomposition of plant tissue and other organic compounds to form humus. The basic processes of horizonation are thought to occur in all types of soils, although they operate with very different relative intensities depending on the five soil-forming factors discussed below. **Haploidization** refers to processes (e.g., pedoturbation) that mix the soil and tend to destroy horizonation (Buol et al. 2011).

4.2 Anthropedogenic Processes

Anthropedogenic processes are defined as artificial conditions which differ from, or significantly modify, the natural processes or factors of soil formation. Human activities often accelerate or disrupt pedogenesis, but many anthropogenic processes can be categorized using the same terminology employed to classify natural soil-forming processes (Table 4.2). On the other hand, some anthropedogenic processes, such as sealing the ground surface with manufactured material (e.g., concrete pavement), have no counterpart in the natural world. Human activities can have direct or indirect impacts, and the effects may be reversible or irreversible. Overall, human impacts on soils vary with geocultural setting, and the technological level of development of any given civilization. Hence, they range from relatively minor impacts at archaeological sites produced by humans using primitive tools or animal labor (e.g., wooden plow), to extensive changes wrought by giant pieces of mechanized equipment used in modern agriculture, mining, and urbanization.

Weathering and horizonation can be affected by human activities. Weathering can be accelerated or retarded by human modifications which affect the drainage characteristics of the soil. For example, irrigation agriculture or the use of a sprinkler system can accelerate leaching and weathering, whereas compaction or construction of a manufactured layer can slow or stop weathering. Horizonation can be changed by human activities which affect additions, losses, translocations and transformations occurring within the soil. Artificial additions can be in the form of natural materials (e.g., animal manure), but man often introduces anthropogenic substances (e.g., waste building materials, coal combustion products) into soils. Losses of chemical matter usually involve human activities that impede, or accelerate those losses which naturally occur in soils. Translocation processes are perhaps most affected by artificial mixing (anthroturbation), ultimately resulting in haploidization. Transformations are strongly affected by compaction, which impedes leaching or aeration, and thus stifles chemical weathering and

Table 4.2 Comparison between natural and anthropogenic soil-forming processes (modified from Buol et al. 2011)

Pedogenic process	Natural example	Anthropogenic example
Additions to the soil		
Enrichment	Plant nutrients added via organic matter decomposition	Chemical elements added via deicing salt runoff, acid rain, fertilizer additions, paint and other building materials
Cumulization	Floodplain deposition; volcanic ash deposition	Deposition of fill at a demolition site; airborne deposition of fly ash
Melanization	Biodecomposition of prairie grasses	Airborne deposition of soot
Littering	Organic litter on forest floor	Artificial mulch used in gardening and landscaping
Losses from the soil		
Erosion	Cut bank of meandering stream	Excavation for building construction, borrow pit, highway, surface mine, etc.
Leaching	Seasonal snowmelt	Irrigation; lawn sprinklers
Translocation within the soil		
Pedoturbation	Ground squirrel burrows	Grading, excavating and backfilling surface mine or demolition site (anthroturbation)
Transformations within the soil		
Mineralization	Release goethite from decomposed humic materials	Release of magnetite from decomposed fly ash
Loosening	Plant root growth; pedoturbation by indigenous earthworm species	Plowing; burrowing by invasive species of earthworms
Hardening	Microbiotic crust formation	Compaction by farm or earthmoving equipment

biodecomposition. Sealing the ground surface with manufactured material can potentially cause all soil-forming processes to completely cease.

Human activities can have significant physical, chemical, mineralogical, and biological impacts on soils (Table 4.3). Human-induced physical alterations to soils can be extensive where surface mining and urbanization are taking place as a result of cut-and-fill operations. In agricultural settings, soils are impacted by the excavation of ditches or irrigation canals, cultivation, plowing, terracing, and compaction caused by trampling by humans or animals, or excessive loading by mechanized or non-mechanized equipment. Water management activities and warfare can also have significant physical impacts on soils. At archaeological sites, soils are impacted physically by the construction of tells, and ceremonial and burial mounds. Agriculture has many chemical impacts on soils. The most common elements added to soils as fertilizers are C, N, P, and Ca with lesser amounts of K, Mg, S, Cu and Zn. Organic matter is also commonly added in the form of manure and other animal wastes, charcoal and biochar, along with organochlorine and

4.2 Anthropedogenic Processes

Table 4.3 Classification and impacts of human activities on soils

Anthropogenic activity	Impacts on soils
Physical	
Excavation	Removal of part, or all, of preexisting soil; exposure of fresh parent material; mixing of earth materials; modified hydrological conditions; destruction of microbial populations
Construction	Mixing of earth materials; burial of soil; modification of hydrological conditions; destruction of microbial populations
Compaction	Decreased aeration, leaching and translocation; reduction of microbial populations
Cultivation, plowing	Mixing of the surface horizon; destruction of soil structure; increased soil erosion
Terracing	Modified hydrological conditions
Dam construction on floodplains	Reduction of sedimentation and leaching; lowering of water table
Construction of artificial drains	Water table drop; increased oxidation and soil structure
Warfare	Increased mixing and compaction; addition of metallic artifacts and bomb-making chemicals; addition of toxic compounds or pathogens from human remains
Chemical	
Fertilizers	Change plant nutrient content and pH
Additions of organic wastes, charcoal, biochar, etc.	Change in composition of organic soil components; increased aeration and water-holding capacity; increased cation exchange capacity
Cropping	Removal of plant nutrients via harvesting; oxidation of soil organic matter
Pesticides	Increased toxic organic compounds or heavy metals; reduction/change microbial population
Deicing salt	Organic matter dissolution; leaching of nutrients and trace metals
Acid rain	Accelerated leaching of plant nutrients
Acid mine drainage	Soil acidification; heavy metal contamination
Human habitation	Increased organic matter and phosphorous content
Mineralogical	
Irrigation	Increase in organic matter and soluble salt content; decreased soil structure
Ponding	Reducing conditions cause dissolution of oxide and other minerals; increased leaching and organic matter content
Fire	Conversion of goethite or hematite to magnetite; alter silicate minerals to glass
Artifacts	Change nutrient concentrations, pH, base saturation, mineralogy, etc.; addition of artificial minerals not present in natural soils; increased toxic heavy metals

(continued)

Table 4.3 (continued)

Anthropogenic activity	Impacts on soils
Biological	
Devegetation	Increased erosion; decreased infiltration, organic matter and nutrients
Domesticated plants and animals	Change in composition of soil organic matter; change in soil biota
Invasive species	Change in composition of soil organic matter; change in soil biota
Pesticides	Change or elimination of soil biota

arsenic-bearing pesticides. Urban soils are often enriched in Pb, As, Hg, Ni, Cr, and various toxic organic compounds as a result of modern technology and industrialization. They are also impacted by acid rain, and fertilizer and deicing salt runoff. Mine-related soils are typically impacted by acid mine drainage and various heavy metals, whereas soils at archaeological sites are perhaps most noticeably enriched in phosphorous and organic matter. In agricultural settings, the mineralogical composition of soils can be affected by the accumulation of soluble salts, or the dissolution of Fe-oxides under reducing conditions (e.g., rice paddy soils). Artifacts, such as coal combustion products and waste building materials, can also contribute minerals of artificial origin (e.g., mullite, portlandite, ettringite) not found in natural soils. Logging and deforestation, cropping, the introduction of domesticated and invasive plant species, and the application of pesticides, can also have major impacts on the biological characteristics of anthropogenic soils.

4.3 Humans as a Soil-Forming Factor

The state factor approach to soil genesis is a theoretical framework in which soil is defined in terms of five soil-forming factors and "mathematically" expressed as:

$$\mathbf{S} = f(\mathbf{c}, \mathbf{l}, \mathbf{o}, \mathbf{p}, \mathbf{t}\ldots) \qquad (4.1)$$

where $\mathbf{S}$ is the whole soil; $\mathbf{c}$ is climate; $\mathbf{l}$ is landscape position (relief, topography); $\mathbf{o}$ is organisms; $\mathbf{p}$ is parent material; and $\mathbf{t}$ is time. The equation has never been solved, but Jenny (1941) studied the variation in one factor while all of the other factors were held constant. Qualitative statements about soils forming as a function of any one factor were called sequences (climosequence, toposequence, biosequence, lithosequence, chronosequence).

Human activities modify one or more of the five soil-forming factors in any given situation (Table 4.4), but some authors have stopped short of designating humans as a sixth factor of soil formation (Bidwell and Hole 1965; Schaetzl and Anderson 2005). Jenny himself considered man to be part of the **o**-factor (Amudson

4.3 Humans as a Soil-Forming Factor

Table 4.4 Impacts of human activities on the five soil-forming factors

Anthropogenic activity	Impacts
Climate	
Agriculture	Increased water by irrigation and rain-making by cloud-seeding; change albedo
Industrialization	Global warming
Landforming	Change soil aspect or drainage characteristics
Urbanization	Change wind and weather patterns; change water content by artificial drainage and sealing
Landscape position (Topography)	
Agriculture	Increased furrowing, soil erosion and fluvial sedimentation; terracing; land leveling; excavation for ditches, canals
Mining/Dredging	Ground subsidence or collapse; land lowering, raising and leveling
Urbanization	Increased soil erosion and fluvial sedimentation; terracing; land leveling; excavation for ditches, canals
Human occupation	Ceremonial and burial mound-building; midden and tell construction
Organisms	
Agriculture	Introducing and controlling plants and animals; devegetation
Urbanization	Introduction of invasive plant and animal species; devegetation
Mining	Devegetation; introduction of invasive plant and animal species; destruction of microbiome
Parent material	
Excavation	Exposure of fresh parent material to weathering
Construction/Demo.	Mixing and fresh exposure of soil parent materials; additions of artifacts
Agriculture	Mixing of soil parent materials; additions and losses of mineral and organic materials
Human occupation	Additions of artifacts, human and animal wastes; exposure or burial of fresh parent material
Urbanization	Additions of coal combustion products and other artifacts
Time	
Excavation	Reset soil-formation clock to time zero
Construction/Demo.	Reset soil-formation clock to time zero; temporarily stop soil-formation clock
Agriculture	Slow soil-formation clock by erosion or excessive removal of nutrients and organic matter; reverse soil-formation clock by artificial ponding

and Jenny 1991). Given that humans are unlike other organisms because they deliberately manipulate soils independently of the other soil-forming factors, and are affecting soils on a much greater scale than other organisms through the use of technology, the collective impacts of humans on soil formation have been regarded collectively by some as a sixth soil-forming factor (Yaalon and Yaron 1966; Effland and Pouyat 1997; Dudal 2005; Leguedois et al. 2016). Thus, Eq. 4.1 may be rewritten as:

$$S = f(\mathbf{a}, \mathbf{c}, \mathbf{l}, \mathbf{o}, \mathbf{p}, \mathbf{t}\ldots) \qquad (4.2)$$

where **a** is the anthropogenic factor. Using the anthropogenic factor, an anthroposequence can be defined as an assemblage of related soils differing in their anthropogenic characteristics, the other soil-forming factors being held constant (Effland and Pouyat 1997).

Humans modify the climate factor (temperature and precipitation) by adding or removing water by irrigation or artificial drainage, respectively, by raising temperature via fire, global warming and the urban heat-island effect, and by sealing soils beneath a manufactured layer such as pavement. The organism factor has been modified by devegetation or deforestation, by the introduction of domesticated plants and animals, through the introduction of invasive species, and by additions of organic matter such as household, human and animal wastes. The landscape position factor has been altered by land raising or leveling, excavation, ground subsidence, and terracing. Warfare is known to have catastrophic effects on local microtopography (Hupy and Schaetzl 2008). The parent material factor is affected by artificial additions of bone, shell, ash and other artifacts, and by the removal of certain materials such as soluble salts. The time factor has been disrupted by burial of soil beneath fill or sealing beneath a manufactured layer, whereas exposure of fresh parent materials by scalping, excavation, or mixing and deposition of fill resets the pedogenic clock to time zero.

4.4 Rates of Anthropogenic Soil Formation

As discussed previously in Chap. 3, it is known from historic records that some metagenic anthrosoils have developed over hundreds or thousands of years depending on how long agriculture has been practiced in any given area. The time required to develop neogenic anthrosols is also relatively well known from studies of mine spoils and urban fills whose ages are constrained by historic records. A-horizons are usually the first feature to develop during neopedogenesis (Simonson 1959). A-horizons a few cm thick have been reported in mine spoils in ≤ 5 years old (Daniels and Amos 1981; Haering et al. 2004), and they may be 5–10 cm thick and contain weak granular structure after ≥ 5 to 20 years (Anderson 1977; Hallberg et al. 1978; Shafer et al. 1980; Wood and Pettry 1989; Varela et al. 1993; Zarin and Johnson 1995). A-horizons are also reported in urban soils developed in demolition site fills within 12–24 years (Howard and Olszewska 2011; Howard et al. 2013). Cambic B-horizons defined by weak structure, a color change, and by rare argillans have been reported in mine soils after 40–50 years (or less) of development (Shafer et al. 1980; Daniels and Amos 1981; Haering et al. 2005). However, in some cases the structure did not persist, or was apparently inherited from the parent material. Weak structure was also observed in 18 year-old demolition site soils which appeared to be inherited from soil parent materials

(Howard and Shuster 2015). Hallberg et al. (1978) did not detect cambic B horizons in 100 year-old spoil in railroad cuts derived from loess, and Parsons et al. (1962) concluded that 2500 years were required to develop prominent B-horizon structure in soils developed on Indian burial mounds. Thus, most anthropogenic soils on recently formed anthropogenic landforms are characterized by A-C profiles. Cambic-like B horizons may be present after several decades, but well developed cambic horizons may require a century or more to form, and argillic horizons, several millennia.

References

Amudson R, Jenny H (1991) The place of humans in the state factor theory of ecosystems and their soils. Soil Sci 151:99–109

Anderson DW (1977) Early stages of soil formation on glacial till mine spoils in a semi-arid climate. Geoderma 19:11–19

Bidwell OW, Hole FD (1965) Man as a factor of soil formation. Soil Sci 99:65–72

Buol SW, Southard RJ, Graham RC, McDaniel PA (2011) Soil genesis and classification. Wiley-Blackwell, West Sussex, UK, 543 p

Daniels WL, Amos DF (1981) Mapping, characterization and genesis of mine soils on a reclamation research area in Wise County, Virginia. Symposium on surface mining hydrology, sedimentology and reclamation. University of Kentucky, Lexington, KY, pp 261–265

Dudal R (2005) The sixth factor of soil formation. Eurasian Soil Sci 38:S60–S65

Effland WR, Pouyat RV (1997) The genesis, classification, and mapping of soils in urban areas. Urban Ecosyst 1:217–228

Fanning DS, Fanning MCB (1989) Soil morphology, genesis, and classification. Wiley, New York, 395 p

Haering KC, Daniels WL, Galbraith JM (2004) Appalachian mine soil morphology and properties: Effects of weathering and mining method. Soil Sci Soc Am J 68:1315–1325

Haering KC, Daniels WL, Galbraith JM (2005) Mapping and classification of southwest Virginia mine soils. Soil Sci Soc Am J 69:463–472

Hallberg GR, Wloonhaupt NC, Miller GA (1978) A century of soil development in spoil derived from loess in Iowa. Soil Sci Soc Am J 42:339–343

Howard JL, Olszewska D (2011) Pedogenesis, geochemical forms of heavy metals, and artifact weathering in an urban soil chronosequence, Detroit, Michigan. Environ Pollut 159:754–761

Howard JL, Shuster WB (2015) Experimental order one soil survey of vacant urban land, Detroit, Michigan. Catena 126:220–230

Howard JL, Dubay BR, Daniels WL (2013) Artifact weathering, anthropogenic microparticles, and lead contamination in urban soils at former demolition sites, Detroit, Michigan. Environ Pollut 179:1–12

Hupy JP, Schaetzl RJ (2008) Soil development on the WWI battlefield of Verdun, France. Geoderma 145:37–49

Jenny H (1941) Factors of soil formation. McGraw-Hill, New York

Leguedois S, Sere G, Auclerc A, Cortet J, Huot H, Ouvrard S, Watteau F, Schwartz C, Morel JL (2016) Modeling pedogenesis of Technosols. Geoderma 262:199–212

Parsons RB, Scholtes WH, Riechen FF (1962) Soils of Indian mounds in northeastern Iowa as benchmarks for studies of soil genesis. Soil Sci. Soc. Am. Proc. 26:491–496

Schaetzl RJ, Anderson S (2005) Soils: genesis and geomorphology. Cambridge University Press, 817 p

Shafer WM, Nielson GA, Nettleton WD (1980) Minesoil genesis and morphology in a spoil chronosequence in Montana. Soil Sci Soc Am J 44:802–808

Simonson RW (1959) Outline of a generalized theory of soil genesis. Soil Sci Soc Am Proc 23:152–156

Varela C, Vazquez C, Gonzalez-Sangregorio MV, Leiros MC, Gil-Sotres F (1993) Chemical and physical properties of opencast lignite minesoils. Soil Sci 156:193–203

Wood CW, Pettry DE (1989) Initial pedogenic progression in a drastically disturbed prime farmland soil. Soil Sci 147:196–207

Yaalon DH, Yaron B (1966) Framework for man-made soil changes—an outline of metapedogenesis. Soil Sci 102:272–277

Zarin DJ, Johnson AH (1995) Base saturation, nutrient cation, and organic matter increases during early pedogenesis on landslide scars in the Luquillo Experimental Forest, Puerto Rico. Geoderma 65:317–330

Chapter 5
Artifacts and Microartifacts in Anthropogenic Soils

Artifacts are objects >2 mm, whereas microartifacts are 0.25–2.0 mm, in size that were produced, modified, or transported from their source, by human activity. Artifacts are typically coal-related wastes (coal, cinders, etc.), waste building materials (brick, mortar, etc.), industrial wastes (coked coal, slag, etc.), and objects of archaeological significance (pottery, bone, etc.). Artifacts can be classified into carbonaceous, calcareous, siliceous, ferruginous and miscellaneous types based on composition. Carbonaceous artifacts are comprised primarily of organic compounds, calcareous types of carbonate minerals, siliceous types of silicate minerals and glass, ferruginous types of Fe-oxides, and miscellaneous types of sulfate minerals, phosphate minerals, etc. Artifacts are often comprised of minerals and glassy substances not found in natural soils, and the effects these phases and their weathering products may have on anthropogenic soil properties are poorly understood. The relative persistence (weathering stability) of artifacts is thought to be: Glass ~ Cinders ~ Glazed brick > Unglazed Brick > Nails > Bone > Concrete > Mortar > Plaster or Drywall ~ Paint. Many microartifacts (MAs) can be identified unequivocally by reflected light microscopy with the same petrographic features used to identify natural microparticles. MAs may be important because even when present in relatively small quantities (<10% by volume) they can have a significant effect on the physical, chemical, and geophysical properties of anthrosoils. Some MAs contain relatively high concentrations of toxic trace elements, hence the recognition of MAs can be important for interpreting the potential adverse environmental impacts of contaminated anthropogenic soils. The effect of MAs on the geophysical properties of anthropogenic soils has been found to be a useful tool for mapping urban soils, and for geospatial analysis of heavy metal contamination.

5.1 Introduction

Artifacts are defined by soil scientists and archaeologists as objects >2 mm in size that were produced, modified, or transported from their source, by human activity (Dunnell and Stein 1989; Sherwood 2001; IUSS Working Group 2015; Soil Survey Staff 2014). Artifacts are usually objects created for some practical use in habitation, manufacturing, excavation or construction activities. However, artifacts can be particles of natural origin (e.g., coal) if they have been transported by humans to a site from a different source location, or if they owe their presence at the Earth's surface to human activity. **Microartifacts** have been defined by archaeologists as artificial objects 0.25–2.0 mm in size. In this book, the terms artifact or macroartifact will be used interchangeably for anthropogenic particles >2 mm in size based on the fact that they are visible with the naked eye. Microartifact refers to particles 0.25–2.0 mm in size, which require microscopic analysis. The term **microparticle** will be used for anthropogenic particles (e.g., fly ash) <0.25 mm in size, and **anthropogenic particle** will be used as a generic term for an artificial particle of any size (Howard and Orlicki 2016).

Anthropogenic particles are typically found as waste materials in anthropogenic soils (Table 5.1). Artifacts at archaeological sites include discarded natural materials that have been altered by human action, such as debitage and simple tools composed of stone, bone and wood. They also include natural materials that are present in an archaeological context only because of human activities, such as seeds, nuts, shells, coal, asphalt, etc. Human-manufactured artifacts include household items such as pottery, glassware, clothing and coins, and very ornate forms of ceremonial jewelry composed of precious metals, gemstones, and perhaps other valuable materials. Artifacts of archaeological significance are also found in certain types of agricultural anthrosoils, mainly Hortic, Plaggic and Pretic Anthrosols, as discussed further in Chap. 7. Mine-related anthrosoils are another special case in which natural materials are considered to be artifacts. In this case, the rock materials comprising mine spoils are present at the Earth's surface only because of blasting, excavation and other human activities. On this basis, mine soils are classified as Technosols, as discussed further in Chap. 9. Anthropogenic particles are ubiquitous in urban soils, where they vary as a function of land use history. Waste building materials are perhaps the most common type, and are particularly widespread in residential and industrial anthrosoils. Glass, brick, and various ceramic artifacts are especially common due to their resistance to weathering. Coal-related wastes are characteristic of industrial urban soils, but they are also common in older residential soils as a legacy from the era of domestic coal use in the 19th and early 20th centuries. Coked coal, other types of coal-related wastes, and various types of slag are common artifacts found in industrial urban soils.

Table 5.1 Common types of anthropogenic particles found in anthropogenic soils

Type of particle		Definition	Source
Coal-related wastes	Coal	An organic sedimentary rock comprised of carbonized plant fossils and trace amounts of mineral matter	Domestic, commercial fuel source
	Coal cinders	The gravel-sized solid waste material left after coal is burned at 1200–1500 °C	Domestic stoves, steam locomotives; power plants; iron smelting
	Coal ash	The sand-sized and finer inorganic solid waste material produced by coal combustion	Incidentally related to coal combustion
Waste building materials	Wood (peat)	The hard fibrous xylem of trees and shrubs	Used in wood-frame construction
	Charcoal	A residue of mainly black carbon produced by the incomplete oxycombustion of wood	Incidentally related to use of wood as a fuel or building material
	Asphaltic concrete	Bituminous cement with rock aggregate	Used primarily as road pavement
	Lime Concrete	A mixture of lime cement and gravel-sized aggregate	Used for road pavement, sidewalks and building construction
	Mortar	A mixture of lime cement and sand-sized aggregate	Used in masonry
	Cinder block	A lime concrete block containing gravel-sized aggregate	Used in masonry to construct walls and foundations
	Soda-lime window glass	A transparent, amorphous mineral-like material formed from silica	Used to make window pane, bottles, etc.
	Ceramic brick	Solid, coherent rock-like blocks produced by firing clay at 900–1100 °C	Used in masonry
	Ceramic pipe, tile, etc.	Solid, coherent rock-like items produced by firing clay at 900–1100 °C	Used for water and sewer pipes
	Corroded iron (nails)	A composite material produced by heating iron ore, limestone and coal to about 1500 °C	Used in wood-frame construction
	Drywall	Large sheets of boards comprised of prehardened plaster of Paris (gypsum)	Used for building interior walls and ceilings

(continued)

Table 5.1 (continued)

Type of particle		Definition	Source
Industrial wastes	Coked coal	A hard, dark gray or black carbonaceous fuel produced by heating coal to 300 °C	Used as a fuel for smelting iron and in coal-fired power plants
	Metalliferous slag	An inorganic, ferruginous, crystalline to non-crystalline waste produced by smelting iron ore, coked coal and limestone at 1500–2000 °C	Iron- and steel-making
	Glass slag	An inorganic, non-crystalline waste produced by smelting iron ore, coked coal and limestone at 1500–2000 °C	Iron- and steel-making
Archaeological materials	Pottery	Household items produced by firing clay at 900–1100 °C	Domestic use
	Bone	An inorganic material comprised primarily of apatite	Primarily the remains of modern farm animals
	Bottle glass	A transparent, non-crystalline mineral-like material	Domestic and commercial uses

5.2 Macroartifacts

Artifacts can be classified into five basic types on the basis of composition (Table 5.2). Carbonaceous artifacts are comprised primarily of organic compounds, calcareous types of carbonate minerals, siliceous types of silicate minerals and glass, ferruginous types of Fe-oxides, and miscellaneous types of sulfate minerals, phosphate minerals, etc. In any given anthrosoil, most of these materials may be found as gravel-sized particles, and in a variety of finer size grades. However, some materials (e.g., coal-ash) are found only as microartifacts or microparticles. Artifacts are often comprised of phase associations, or type of minerals or mineraloids, not found in nature, and the effects these phases and their weathering products may have on anthropogenic soils are poorly understood. Note that the term "charcoal" is used here for the charred remains of wood produced by oxycombustion (Table 5.1), as opposed to "biochar," which is produced by low temperature (<700 °C) thermal decomposition with a limited supply of O_2 (Barrow 2012). Anthropogenic particles produced by iron smelting are denoted as "metalliferous slag" and "glass slag." Coal combustion produces anthropogenic particles which have also been referred to as slag, but here they are called "cinder" to avoid ambiguity. The term "concrete" refers to a lime-based material unless otherwise indicated as asphaltic concrete.

5.2.1 Carbonaceous Artifacts

Carbonaceous artifacts in anthropogenic soils are primarily wood, charcoal, coal, coke, and asphalt (Table 5.2). They are generally composed of a complex assemblage of organic compounds which are converted to black carbon and condensed aromatic organic compounds upon combustion. Wood is commonly used as a building material (lumber), and is sometimes found as tools and posts at archaeological sites. Lumber is produced from the xylem of trees which has a laminated cellular structure comprised primarily of cellulose and lignin. Wood undergoes partial thermal decomposition at ~250 to 400 °C. Thus, charcoal is comprised of hydrocarbons and black carbon (Rutherford et al. 2005; Brodowski et al. 2005), mixed with remnant laminar structures of parent plant materials (Cornelissen et al. 2005; Forbes et al. 2006). Coal is comprised of vitrinite macerals (Fig. 5.1a), which also retain the laminar microstructure of wood (Petrakis and Grandy 1980; ICCP 1998; Suarez-Ruiz 2012).

The coking (carbonization) process involves heating certain types of bituminous coals to a temperature of ~1100 °C (or more) in the absence of oxygen to form a hard spongy mass of nearly pure carbon (Suarez-Ruiz 2012). As the coal is heated to 350–450 °C, it softens into a plastic mass, devolatilizes and vesiculates, often with considerable foaming. As it is further heated to 450–550 °C, it begins to harden into coke (Gray 1991). Hence, coked coal is characteristically vesicular.

Table 5.2 Composition of anthropogenic particles commonly found in anthropogenic soils

Type of artifact		Composition	References
Carbonaceous	Wood (lumber)	A natural composite of cellulose and hemi-cellulose fibers embedded in a matrix of lignin	Singh et al. (2010), Fromm (2013)
	Charcoal	Aromatic hydrocarbons, black carbon, some remnant lignin	Brodowski et al. (2005), Forbes et al. (2006)
	Coal	Organic macerals of various types depending on rank, and minor quartz, illite, kaolinite, feldspar, calcite, dolomite, pyrite, galena	Petrakis and Grandy (1980), ICCP (1998), Ward (2002), Suarez-Ruiz (2012), Howard and Orlicki (2016)
	Coked coal	Inertinite macerals and black carbon	Gray (1991), Choudhury et al. (2008)
	Asphaltic concrete	Bituminous hydrocarbons	Yang et al. (2010b), Brown (2013), Howard and Orlicki (2016)
Calcareous	Lime concrete	Various mixtures of calcite, portlandite, belite, alite, tobermorite, ettringite, etc. with variable rock and mineral fragment types	Kosmatka et al. (2002), Lane, 2004, Van Oss (2005), Howard and Orlicki (2016)
	Mortar	Various mixtures of calcite, portlandite, belite, alite, tobermorite, ettringite, etc. with variable rock and mineral fragment types	Kosmatka et al. (2002), Van Oss (2005), Howard and Orlicki (2016)
	Cinder block	Similar composition to concrete, but containing blast furnace slag and possibly fly ash	Howard and Orlicki (2016)
	Lime brick	Various mixtures of calcite, portlandite, belite, alite, tobermorite, ettringite, etc. with variable rock and mineral fragment types	Howard and Orlicki (2016)
Siliceous	Coal cinders	Aluminosilicate glass, mullite, quartz, magnetite, hematite	Ward and French (2005), Howard and Orlicki (2016)
	Coal ash	Aluminosilicate glass, mullite, quartz, magnetite, hematite, calcite, gypsum	Fisher et al. (1976), Carlson and Adriano (1993), Ward and French (2005), Lanteigne et al. (2012)

(continued)

5.2 Macroartifacts

Table 5.2 (continued)

Type of artifact		Composition	References
	Burnt shale	Quartz, glass, mullite, hematite	Howard and Orlicki (2016)
	Ceramic brick	Aluminosilicate glass, mullite, wollastonite, cristobalite, sanidine, hematite, quartz	Livingston et al. (1998), Cultrone et al. (2004, 2005), Howard and Orlicki (2016)
	Ceramic pipe, tiles, etc.	Aluminosilicate glass, mullite, wollastonite, cristobalite, sanidine, hematite	Stoltman (2001), Reedy (2008), Rapp (2009), Howard and Orlicki (2016)
	Pottery sherds	Aluminosilicate glass, phyllosilicates, mullite, wollastonite, cristobalite, sanidine, hematite	Rapp (2009)
	Soda-lime glass	Amorphous silica, usually containing Na, Ca, and coloring agents such as Fe, Cu or Co	Rapp (2009), Mukherjee (2011)
	Glass slag	Glass, hematite, magnetite	Fredericci et al. (2000), Howard and Orlicki (2016)
Ferruginous	Ferruginous slag	Glass, merwinite, melilite, wollastonite, belite, olivine, wustite, magnetite, hematite, calcite, portlandite	Bayless et al. (2004), Yildirim and Prezzi (2011), Piatak and Seal (2012), Howard and Orlicki (2016)
	Corroded iron (nails, etc.)	Ferrite, ferrihydrite and goethite	Asami and Kikuchi 2002, Neff et al. (2005), Howard et al. (2013a), Howard and Orlicki (2016)
	Iron microspheres	Magnetite	Lanteigne et al. (2012), Howard and Orlicki (2016)
Other	Drywall	Gypsum possibly containing plastic fibers or animal hairs	Diamant (1970), Howard and Orlicki (2016)
	Bone	Collagen fibers and hydroxyapatite	Berna et al. (2004), Howard and Orlicki (2016)

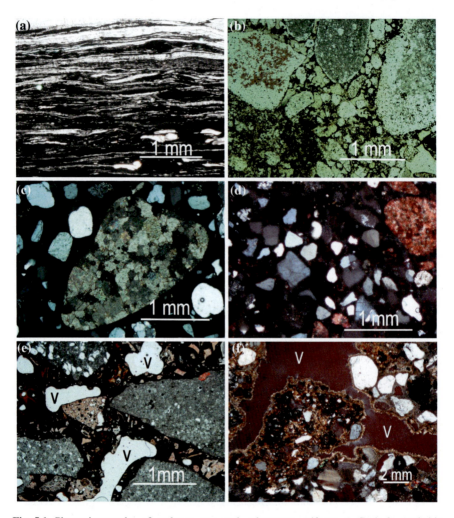

Fig. 5.1 Photomicrographs of carbonaceous and calcareous artifacts: **a** Coal; **b** Asphaltic concrete; **c** Concrete; **d** Mortar; **e** Cinder block; **f** Voids in weathered mortar

Coked coal was widely used in iron-making blast furnaces, and other metal smelting operations. It was produced widely until the 1970s as a by-product of manufactured gas operations.

Asphaltic concrete is a composite construction material comprised of gravel-sized, mineral-based aggregate held together by a viscous binder (asphalt mastic). The mastic is composed of fine sand and asphalt binder which holds the larger aggregate together. The asphalt binder is a black, highly viscous, carbonaceous liquid formed by boiling petroleum at ∼525 °C (Roberts et al. 1994). Rolled asphaltic concrete pavement is produced by mixing asphalt and aggregate at ∼150 °C, and compressing layers on the road bed with a heavy mechanical roller.

Asphaltic concrete has been used mainly for road pavement in the United States since the 1870s (Brown 2013). It is typically comprised of 5% asphaltic binder and 95% aggregate (Fig. 5.1b).

Wood is a non-persistent artifact due to its tendency to biodecompose. However, it can be well preserved under reducing conditions in waterlogged soils, or in highly compacted, poorly aerated, urban soils. Coal, coal coked coal, asphalt and charcoal are very persistent artifacts. They are known to be highly recalcitrant to chemical and biological degradation, based on their preservation in the geological record, and ^{14}C dates indicating a mean residence time in soils of >1000 years (Krull and Skjemstad 2002). Coal and asphalt are resistant to chemical weathering because of the strongly hydrophobic nature of bitumen, but coal weathers very slowly by oxidation (Chang and Berner 1999), whereas asphalt is essentially unaffected. There is also a growing body of evidence suggesting that significant oxidative degradation of coal can occur, under certain conditions, over a much shorter timescale. Although *unweathered* coal is highly recalcitrant, microbial production of alkaline substances, and/or chelating agents and surfactants, can solubilize *weathered* coal under basic (pH 7–10) conditions (Martinez and Escobar 1995; Hofrichter and Fakoussa 2001). The bacteria (e.g. *Bacillus, Pseudomonas*), actinobacteria (e.g. *Streptomyces*), and fungi (e.g. *Penicillium, Aspergillus*) known to be capable of solubilizing coal are species which are common in the rhizosphere. Extensive studies of biochar and charcoal suggest that black carbon is also more susceptible to biodegradation than previously thought. Current thinking involves a two-pool model in which one black carbon pool degrades on a timescale of decades, whereas the other requires centuries (Nguyen et al. 2008). The weathering characteristics of soot are poorly known, but presumably similar to those of other artifacts comprised of black carbon. Dissolved organic carbon has been widely observed in aquifers and attributed to weathering of black carbon.

5.2.2 Calcareous Artifacts

Concrete, mortar, cinder block, lime brick and lime plaster are the main types of calcareous artifacts found in anthropogenic soils (Table 5.2). Concrete is an artificial rudite-like building material comprised of gravel- and sand-sized rock and mineral fragments bound together as a rigid mass by lime-based cement (Fig. 5.1c). Most concretes are comprised of hydraulic cement, i.e. one that hardens by reaction with water (Van Oss 2005). Concrete is typically $\sim$30 to 60% gravel, 25–30% sand, and 10–15% cement (Kosmatka et al. 2002). Concrete is mixed onsite, and poured as a slurry to form a slab for use as a house foundation, or as a sidewalk or road pavement. It is also fashioned into bricks or blocks for use as masonry. Two types of cement were used in England and the United States during the 19th century prior to the widespread adoption of Portland cement. Ordinary lime (CaO) cement was produced by calcination of pure limestone. It hardened by a combination of hydration and carbonation. An early type of hydraulic cement (called natural cement) was produced by calcination of clay-rich limestone. The earlier types were later replaced in the early 20th century by Portland cements, which are blends of finely ground Portland cement clinker, gypsum or anhydrite, and perhaps other raw materials (Van Oss 2005). Natural cement is comprised primarily of portlandite (Ca(OH)$_2$), belite (calcium silicate) and calcite, whereas Portland cement has a complex composition involving a number of artificially produced silicate (e.g., alite) and sulfate (e.g., ettringite) phases. Mortars are similar to concrete (Fig. 5.1d), but contain only sand-sized aggregate, and are mixed in such a way as to be more plastic and workable. Cinder blocks are a type of concrete masonry unit containing blast furnace cinders as aggregate, and characterized by very large voids (Fig. 5.1e). Mortar was sometimes used to make a type of brick simply called lime brick.

Lime plaster is a type of plaster composed of sand, water, and lime (CaO). It sets and hardens by carbonation, i.e. the relatively slow reaction of CaO with CO_2 from air to form $CaCO_3$. It can remain soft for weeks or months, but once hard it is very durable. Some plaster Roman walls have survived for more than 2000 years under a Mediterranean climate.

$$CaO + H_2O \rightarrow Ca(OH)_2 \qquad (5.1)$$

$$Ca(OH)_2 + CO_2 \rightarrow CaCO_3 + H_2O \qquad (5.2)$$

Concrete artifacts comprised of Portland cement are very persistent in soil even under humid climatic conditions, but 19th century mortar that was comprised of natural hydraulic cement was found to be severely weathered in urban soils (Fig. 5.2) less than 100 years old (Howard et al. 2013). Petrographic analysis showed the presence of large voids in weathered mortar clasts formed by solution of cement binder (Fig. 5.1f), and possibly indicating weathering accelerated by deicing salts or gypsum from decomposing plaster or drywall. Some urban soils showed evidence of accelerated weathering with mortar clasts disintegrating in

Fig. 5.2 Weathered brick and mortar in 100 year-old demolition site anthrosoil

5.2 Macroartifacts

<20 years (Howard and Olszewska 2011). Lime bricks were characterized by very poor durability, and have been banned from use as a building material in many places.

5.2.3 Siliceous Artifacts

Siliceous macroartifacts found in anthrosoils include coal-related wastes such as coal cinders, carbonaceous shale and burnt shale, bricks and other ceramic building materials, pottery, and various types of glass (Table 5.2). Coal cinders (aka as clinker or slag) are inorganic wastes produced by coal combustion at temperatures of ~ 1200 to 1500 °C. Although they are perhaps most commonly 2–10 mm, they can be found as gravel-sized clasts 5 cm or more, in size. In hand specimen, coal cinders may be similar in appearance to metallurgical slag, and both can resemble basaltic cinders. Coal cinders are formed when ash particles, derived from the inorganic constituents of coal, fuse together while degassing in a plastic or molten state. Cinders can also form from fragments of carbonaceous or argillaceous shale associated with coal, which were collected inadvertently during mining, and burned with it. In smaller boiler furnaces (e.g., steam engines), coal cinders are removed from the firebox by a fireman using a long poker, whereas large blast furnaces may be designed so that molten slag can flow out through a hole in the side or bottom. Cinders are generally formed from minerals which are minor constituents of coal, mainly calcite, clay minerals (kaolinite, illite), feldspar, quartz and pyrite (Ward 2002). At a temperature of ~ 900 to 1200 °C, calcite decomposes into lime, and clay minerals and feldspar decompose into glass or high temperature phases such as spinel, mullite, or sanidine. Quartz is stable to 1800 °C and is generally unreactive, but a small amount may melt to form glass or tridymite (Ward 2002). At temperatures of 200–1000 °C, pyrite decomposes to form hematite (Waanders et al. 2003; Hu et al. 2006), which is stable to ~ 1388 °C (Abdullah and Atherton 1964). However, hematite can be converted into magnetite by oxidation at 900–1000 °C (Hu et al. 2006), or rapidly (<3 min) at ~ 600 °C by reduction with only a small amount H_2 (Matthews 1976; Kontny and Dietl 2002; Wagner et al. 2006; Bhargava et al. 2009). The average composition of a coal cinder is $\sim 60\%$ glass, 25% mullite, 8% quartz, 5% iron oxides, and 2% miscellaneous (Ward and French 2005). In thin section, coal cinders are characterized by a vesiculated microtexture, fluxion structure (Fig. 5.3a), and masses of acicular mullite crystals (Fig. 5.3b).

Carbonaceous shale is an argillaceous rock composed primarily of silicates, and containing variable amounts of bituminous materials. Carbonaceous or argillaceous shale are frequently interstratified with coal in coal-bearing strata, and prior to the development of "coal washers" in the 1970s, were collected unavoidably with coal during the mining process. Burnt shale is produced when these materials are accidently burned with the coal. It has a unique mineralogy, including glass, mullite and other high-temperature silicates, produced from the detrital components of

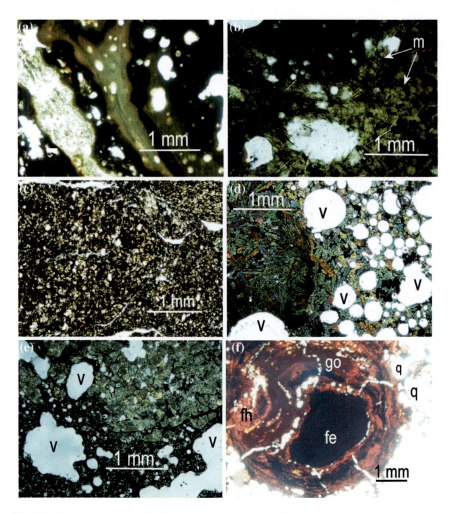

Fig. 5.3 Photomicrographs of siliceous and ferruginous artifacts: **a** Coal cinder; **b** Coal cinder showing acicular mullite crystals (m); **c** Brick; **d** Hornblende-rich steel-making slag; **e** Olivine-rich steel-making slag; **f** Severely corroded hand-forged 19th century square nail showing ferrite core (fe), goethite (go), ferrihydrite (fh), and quartz

shale by coal combustion (Howard and Orlicki 2016). Thus, burnt shale is commonly associated with coal and coal cinder artifacts in anthropogenic soils.

A ceramic brick is an artificial rock-like block produced by firing clay, usually in the range of ~800 to 1200 °C. Ceramic bricks have been used for building construction since ancient times (Rapp 2009). The compositions of bricks depend on historical changes in methods of manufacture as well as variable factors such as the mineralogical composition of clay, the nature of any additives, drying time, and firing temperature and atmosphere (oxidizing vs. reducing). Bricks made in the U.S. during the 19th century were produced by mixing clay with sand (to reduce shrinkage and cracking) and water, shaping and cutting the bricks by hand (or using a mechanical press), and firing at <1000 °C in the ground or in a kiln. Such bricks are typically unglazed and of variable size. Bricks produced since ~1920s, and fired at >1000 °C, are generally glazed, of uniform size and, more recently, hollow or perforated. Bricks fired at <1000 °C generally lack mullite and cristobalite, and tend to be more porous and weak (Ahmad et al. 2008). As the firing temperature increases, carbonates decompose at ~820 to 890 °C, and above ~1000 °C, phyllosilicates decompose to form aluminsilicate glass and hematite. Other high temperatures phases also appear such as mullite, sanidine and cristobalite (Livingston et al. 1998; Cultrone et al. 2004, 2005). During firing, the exterior of the brick develops a glassy coating, while the interior may remain more porous and earthy (Fig. 5.3c). Calcareous bricks tend to have yellow hues, whereas ferruginous components produce red bricks under oxidizing conditions. Bricks fired under reducing conditions are purple, brown or gray (Reedy 2008).

Other ceramic materials, used as building materials, ornamentation, pottery, etc. include earthenware (terracotta), stoneware and porcelain (Rapp 2009). Earthenware refers to the more or less unvitrified, unglazed, coarse, porous, red or reddish-white ceramic material which results from firing at low temperature (<900 °C). Stoneware is fired at higher temperature (1100–1300 °C) and is more vitrified, and thus stronger and less permeable. It often has a glaze which is formed by coating the clay with powered glass before firing (Reedy 2008). Porcelain is a glossy white, hard, dense, impermeable ceramic formed from kaolinite and a special feldspathic glaze by prolonged (~90 h) firing at ~1100 °C, which causes the extensive development of mullite (Duval et al. 2008).

Soda-lime glass is a solid, non-crystalline material, produced by the quenching of silicate melt. Artificial glass has been produced for at least 9000 years (Rapp 2009). Container and window glass are usually manufactured from very pure quartz sand, often by adding sodium (Na_2O) and lime (CaO) as fluxing agents to reduce the melting temperature (Mukherjee 2011). Artificial glass shards sometimes may be distinguished microscopically by pigmentation and colors (e.g., cobalt blue) not normally seen in naturally formed minerals. Writing on container glass from the 19th century sometimes indicates a time or place of origin, and the evolution of bottle-making is such that intact bottle artifacts can often be dated precisely (Hunt 1953).

5.2.4 Ferruginous Artifacts

Ferruginous macroartifacts found in anthrosoils are mainly various objects composed of wrought iron or steel, and the metalliferous slag waste materials associated with their production (Table 5.2). Metallurgical slag is a more or less vitreous, inorganic by-product of the smelting of iron ore during the manufacturing of iron or steel. Iron-making typically involves heating iron ore, coked coal, and a fluxing agent (limestone or dolostone) at ~2000 °C in a heat-resistant, brick-lined blast furnace (Proctor et al. 2000). This generates a molten mass containing lime which reacts with silica to form a molten silicate slag. The slag, which floats on the molten iron, is separated by decanting into slow-moving train cars. The semi-molten slag is then carried away from the furnace where it is dumped forming large piles of rubble. These slag piles are comprised mainly of gravel-sized particles that range from variably colored chunks of glass, to light or dark gray clasts with vesicular texture similar in appearance to basaltic cinders. Steel-making is a similar process, but involves the use of basic oxygen or electric arc furnaces (Yildirim and Prezzi 2011). Both iron- and steel-making slags are usually comprised of variable amounts of metal oxides (or elemental metal) and carbonates, in a matrix of non-crystalline or crystalline silicate material (Bayless et al. 2004; Yildirim and Prezzi 2011; Piatak and Seal 2012). The petrographic characteristics of cindery iron-smelting slag are somewhat like those of basalt (Fig. 5.3d, e), but plagioclase is conspicuously absent. The slag cinders are dense, highly magnetic, and characterized by a distinctive artificial assemblage of phases including glass, magnetite, merwinite, melilite, belite, etc. Slag appears to be a highly persistent artifact because of its glassy composition.

A wide variety of ferruginous macroartifacts comprised of iron or steel is commonly found in anthrosoils including nails, wire, railroad spikes, brackets, rods and other car parts, etc. along with metallic shavings from their manufacture. These objects are typically more or less corroded. Soils containing demolition debris from the 19th century may contain corroded hand-forged wrought iron nails. Wrought iron nails are often encased in a mass of soil particles formed as a result of pedocementation (Fig. 5.4). These weathered nails typically consist of a remnant ferrite nail core (Fig. 5.3f) surrounded by ferrihydrite and goethite, which have locally cemented together the surrounding soil particles (Howard et al. 2013). The corrosion of iron is generally a microbially-mediated electrochemical process that produces variable amounts of ferrihydrite and goethite, depending on local redox

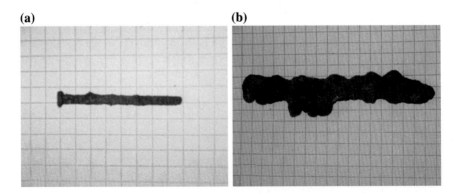

Fig. 5.4 Iron-oxide pedocementation of anthrosoil: **a** Relatively uncorroded hand-forged 19th century square nail; **b** Severely corroded iron nail encased in iron oxide-cemented soil material. Grid spacing 4 mm

conditions (Schwertmann and Fisher 1973; Heitz 1996). Corroded steel structures (Asami and Kikuchi 2002), and iron artifacts at archaeological sites (Neff et al. 2005), are often comprised of goethite and ferrihydrite. These corroded ferruginous artifacts can sometimes be dated using the radiocarbon method.

5.2.5 Miscellaneous Artifacts

Miscellaneous artifacts sometimes found in anthrosoils are bone, gypsum-based plaster and drywall, and paint (Table 5.2). Bones of humans and livestock are typically found in an archaeological context (habitation sites, cemeteries and burial mounds), but they are also found in an urban setting at sites where buildings were demolished that pre-date modern legal regulations banning farm animals within city limits. Bone consists mainly of hydroxyapatite crystals bound together by a smaller amount ($\sim 20\%$) of proteinaceous collagen fibers. Mammalian bones are comprised partly of a spongy, porous tissue, and partly a denser non-porous type. Bones generally weather physically in soil by cracking and exfoliation, and then by increasing chemical dissolution with decreasing pH (Berna et al. 2004). The persistence of bone artifacts varies with burial and climatic conditions. Bones artifacts can dissolve away under acid or strongly leaching conditions in less than 10 years (Schiffer 1987). However, they can persist in calcareous soils under an udic moisture regime for many decades, and under xeric conditions for centuries.

Although the earliest known plasters were lime-based, the use of gypsum-based plasters can be traced back to ancient buildings in the Middle East, China and India constructed before 5000 B.C. Gypsum-based plasters are produced by heating gypsum ($CaSO_4$–$2H_2O$) at ~ 150 °C, causing it to dehydrate. When subsequently mixed with water it rehydrates and hardens as it dries. Ca-aluminosilicate can be added to hasten the hardening process. This form of hydraulic plaster was used extensively by the ancients Romans ~ 400 B.C. who employed mixtures of sand, lime, gypsum and volcanic ash. Plaster was widely used in Europe during the Middle Ages, but hydraulic plaster was not rediscovered until the 18th century. During the 19th and 20th centuries, walls were generally constructed with three layers of gypsum plaster laid over lath-boards nailed end-to-end over a wood frame. However, the use of gypsum boards in building construction began during the late 19th century, after the invention of "Sackett Board," i.e. a wallboard comprised of thin layers of plaster of Paris lying between wool felt paper. Sackett Board was often used as a replacement for wood and as a base for the application of plaster. U.S. Gypsum invented their version of wallboard (drywall) in 1916 which increased greatly in use after the end of WWII. By 1955, roughly 50 percent of new homes were built using gypsum drywall, and the rest were built with gypsum lath and plaster.

Paint is basically a liquid containing various additives and pigmenting agents that is applied to a substrate in thin layers, and then converts to a solid film upon drying. Its origin can be traced back to prehistoric times, and paint was widely used by many ancient civilizations to protect and color the walls, ceilings and floors of buildings. More modern paints used either oil or water as a base, with the pigment and base mixture being ground into a paste with a mortar and pestle. Fillers were a special type of pigment added to thicken the film, and increase the volume of the paint. They were usually made using cheap and inert materials, such as diatomaceous earth, talc, lime, clay, etc. White lead (cerussite) pigments were widely used in the U.S. during the late 19th and early 20th century before being banned for residential use in 1978 by the USEPA and Consumer Product Safety Commission.

Plaster and paint artifacts are typically are laminated and platy. Gypsum ($Ksp = 10^{-4.7}$) and cerussite ($Ksp = 10^{-4.65}$), comprising plaster (drywall) and paint, respectively, are relatively soluble. Hence, plaster and paint are relatively non-persistent artifacts. Paint generally appeared to have weathered away after ~24 years, and drywall after ~30 years, in calcareous urban soils under an udic moisture regime. Similarly, 19th century mortar thought to be comprised of natural hydraulic cement was severely weathered, whereas 20th century concrete made from Portland cement was well preserved, after about 39 years of pedogenesis (Howard et al. 2013). Bone (apatite $Ksp = 10^{-10.2}$) was found to be well preserved after spending nearly 100 years in a demolition site soil in the same area (Howard et al. 2015). Overall, the artifact weathering stability sequence observed in urban soils was: Glass = Cinders = Glazed brick > Unglazed Brick > Nails > Bone > Concrete > Mortar > Plaster or Drywall = Paint (Howard and Olszewska 2011; Howard et al. 2013).

5.3 Microartifacts

Microartifacts (MAs) have been found to be a common component of some urban anthropogenic soils (Howard and Orlicki 2016). They may be important because even when present in relatively small quantities (<10% by volume) they can have a significant effect on the physical, chemical, and geophysical properties of anthrosoils (Howard and Orlicki 2015; Lukasik et al. 2015). Furthermore, some anthropogenic particles of great significance (e.g., coal ash) are found only as MAs. Some MAs contain relatively high concentrations of toxic trace elements, hence the recognition of MAs can be important for interpreting the potential adverse environmental impacts of contaminated anthropogenic soils. The effect of MAs on the geophysical properties of anthropogenic soils has been found to be a useful tool for mapping urban soils (Howard et al. 2016), and for geospatial analysis of heavy metal contamination (e.g., Yang et al. 2010, 2012).

5.3.1 Optical Characteristics

MAs can be identified by reflected light microscopy using the same types of petrographic features used to identify natural microparticles (Howard and Orlicki

5.3 Microartifacts

Micromorphology	Sample	Color	Clarity	Luster	Fracture	Texture	Shape
	**Carbonaceous**						
	Wood (lumber)	light to dark brown and black	opaque	earthy	splintery to fibrous	agranular	angular platy to rod-like and fibrous
	Charcoal	black	opaque	bright vitreous	splintery to fibrous	agranular	angular blocky to prismatic
	Coal	black	opaque	bright vitreous	splintery to hacky jagged and conchoidal	agranular, possible microlamination	angular to very angular blocky
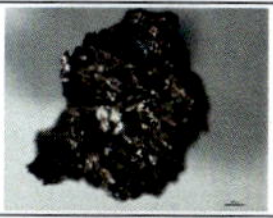	Coked coal	black	opaque	dull to bright some vitreous; iridescence	jagged hacky	agranular vesicular	very angular blocky
	Asphaltic concrete	black	opaque	earthy to resinous	smooth hacky	polymictic granular, aggregatic	subangular to subrounded equant
	**Calcareous**						
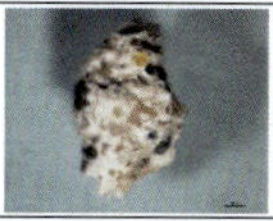	Lime concrete	white	opaque	earthy	smooth hacky	polymictic granular aggregatic	subangular to subrounded equant

Fig. 5.5 Optical characteristic of microartifacts commonly found in anthropogenic soils (Modified from Howard and Orlicki 2016)

2016). This includes such properties as color, clarity, luster, fracture, microtexture and shape (Fig. 5.5). When crushed, composite materials that contain aggregate (asphaltic concrete, mortar, cinder block, etc.) will disarticulate and release natural mineral or rock grains that are indistinguishable from those of natural origin. However, such composite materials can usually be identified as anthropogenic using MAs containing binder materials (asphalt, lime cement, etc.)

Wood MAs (lumber) lack luster and are brown, fibrous or platy with a splintery cleavage. Charcoal MAs retain the fibrous or platy shape and splintery cleavage of wood, but are black with a bright vitreous luster. The lamellar microstructure of wood is partially retained by coal. Hence, some of the coal MAs, with a fibrous or platy morphology and splintery cleavage, are indistinguishable from charcoal (Fig. 5.5). However, coal MAs are more often angular blocky grains with a well developed conchoidal fracture, and may be distinguished by microlamination not seen in charcoal. Coked coal MAs are distinguished from charcoal and coal by their highly vesicular morphology, which contrasts sharply with the splintery fracture of coal and charred wood. Some coke grains are iridescent, possibly due to thin microlamellar coatings of glass, or an unknown crystalline organic compound.

	Lime brick	grayish-white	opaque	earthy	smooth hacky	polymictic granular aggregatic	subangular to subrounded equant
	Mortar 1	white	opaque	earthy	smooth hacky	polymictic granular aggregatic	subangular to subrounded equant
	Cinder block	grayish-white to dark gray	opaque	earthy	jagged hacky	polymictic granular aggregatic vesicular	angular blocky
				Siliceous			
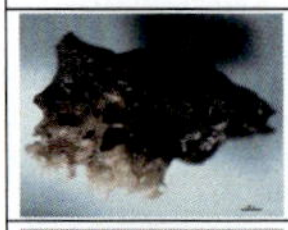	Coal cinders	black to pale brown	opaque to semitranslucent	dull to bright vitreous	conchoidal to jagged hacky	agranular vesicular	angular to very angulara blocky
	Coal ash (microspheres)	gray	translucent	bright vitreous	none	granular	Very well rounded
	Coal ash (agglomerate)	gray to grayish-brown	semi-translucent	dull vitreous	none	granular aggregatic	Well rounded to subangular

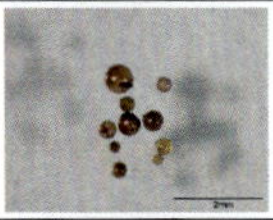

Fig. 5.5 (continued)

Asphaltic MAs containing mastic are black and opaque like coal and coke, but differ in being equant and subrounded with a dull earthy or resinous luster, and a granular, polymictic, aggregate microtexture.

Sand-sized grains of concrete, mortar and lime-brick are generally optically indistinguishable from each other (Fig. 5.5), but MAs containing anthropogenic lime cement differ from most carbonaceous MAs in being light-colored with a polymictic granular texture. Asphaltic concrete has a granular aggregate texture, but is darker and rounder than the lime-based MAs. Cinder block MAs are distinguished by a characteristic gray color, granular vesicular texture, and the presence of highly vesicular pumice-like pieces of blast furnace slag.

Coal cinder MAs are distinguishable from coal by their vesicular microtexture, and from coke by their color (multicolored black and pale brown to greenish-brown), dull vitreous luster, and conchoidal fracture (Fig. 5.5). They are also readily identified by their moderately to strongly magnetic character. Coal ash is comprised of a variety of different particle types, which can be classified generally as either spherical or non-spherical. Glassy, translucent microspheres are common, along with light to dark-colored non-spherical, highly vesicular, pumice-like grains. Microagglomerate is comprised of an agglutination of non-spherical or spherical microparticles. Although some coal-ash particles are in the size range of microartifacts (0.25–2.0 mm), most are <0.25 mm in size. These particles are generally 15 to 150 µm in size (or smaller), and thus prone to eolian transport. Coal ash

5.3 Microartifacts

	Burnt shale	pale pinkish gray to pinkish brown	opaque	Dull vitreous	platy	agranular	subangular to very angular
	Red brick	dark brownish orange to reddish brown	opaque	earthy	smooth hacky	polymictic granular aggregatic	subrounded equant to blocky
	Orange brick	brownish orange	opaque	earthy	smooth hacky	polymictic granular aggregatic	subangular to subrounded equant
	Terracotta	reddish orange	opaque	earthy	jagged hacky	polymictic granular aggregatic	very angular to subangular blocky
	Glazed ceramic	light grayish brown to brownish orange	opaque	earthy to vitreous	Smooth hacky	polymictic granular aggregatic	angular to subrounded blocky
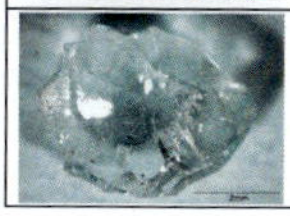	Glass	variable	transparent	bright vitreous	conchoidal	agranular	very angular blocky

Fig. 5.5 (continued)

microparticles are morphologically complex, but various types of microspheres are characteristic. Microspheres are produced as the coal is heated above the softening temperature of 350–450 °C. The coal particles become plastic, and dehydration of clay or thermal decomposition of calcite produces H_2O and CO_2 gas, respectively. At high temperatures, during the coking process, gasification, or combustion, degassing produces bubble-shaped spheres comprised of glass, mineral, or carbonaceous materials by bursting and agglomeration (Fisher et al. 1976; Smith et al. 1979; Gray 1991). Hot gases escaping from the furnace carry these microparticles with them where they may be subsequently deposited in the environment as fly ash. Cenospheres (hollow microspheres) and pleurospheres (hollow microspheres containing smaller microspheres) are the most common component, but coal ash microparticles can be non-spherical, or agglomerates comprised of a mixture of spherical and non-spherical types (Fisher et al. 1976, 1978; Smith et al. 1979). Coal ash is basically comprised of three constituents (Ward and French 2005; Cornellisen et al. 2005): (1) non-crystalline aluminosilicate glass, (2) mineral material (mainly quartz, mullite, spinel, tridymite, sanidine, magnetite, hematite), and (3) carbonaceous material derived from unburnt coal, or by condensation of volatilized organic compounds (soot). Coal ash microparticles generally range in composition from siliceous to ferruginous and carbonaceous (Howard et al. 2013, 2015; Howard and Orlicki 2016). However, microparticles with similar micromorphology may have different elemental compositions (Fisher et al. 1978). Glassy siliceous microspheres

	Glass slag	dark to pale green; variable	opaque to translucent	bright vitreous	conchoidal	agranular	very angular blocky
	Ferruginous						
	Wrought iron (corroded)	orange to dark brown	opaque	earthy			
	Ferruginous slag	very dark gray	opaque	Dull resinous to bright metallic	jagged hacky	agranular vesicular	angular to very angular blocky
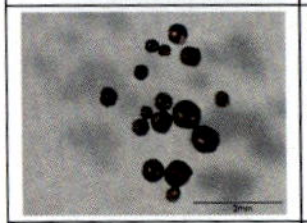	Coal ash (microspheres)	brown, red and black	opaque	metallic	none	granular	Very well rounded
	Miscellaneous						
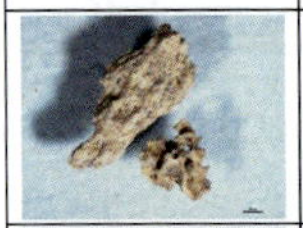	Bone	light brownish yellow to yellowish brown	opaque	earthy	jagged to smooth hacky	agranular porous	angular to very angular blocky to platy
	Drywall	white	opaque	earthy	smooth hacky	agranular vesicular	subrounded equant

Fig. 5.5 (continued)

tend to be translucent (Fig. 5.5), whereas ferruginous types are typically dark colored and opaque. Ferruginous microspheres are commonly comprised of magnetite (Lu et al. 2011; Lanteigne et al. 2012; Minyuk et al. 2013), formed by high temperature alteration of pyrite and hematite. Carbonaceous microparticles (soot) can be spherical, irregularly shaped, or present as in inclusions in siliceous grains. Many fly ash particles are agglomerates comprised of an agglutination of microparticle types including Fe-rich, magnetic aluminosilicate glass.

Ceramic brick MAs (Fig. 5.5) have an earthy luster and granular microaggregatic texture, but are generally reddish or orangish brown in color, and contain finer grained aggregate than mortar. Terracotta and stoneware MAs, including pottery, are microscopically indistinguishable from brick MAs. Glazed ceramic pipe MAs are also similar to brick, but some grains have a slightly more vitreous luster. They are also characterized by an association of gray and reddish brown particles, the later often coated with a conspicuous glassy glaze. Window and bottle glass MAs have extreme angularity, conchoidal fracture, and are differentiated from quartz by their characteristic transparency. Glass slag MAs are similar to manufactured glass shards, but tend to be less transparent. They range from darker green and opaque grains with a dull vitreous or resinous luster, to pale green translucent grains with a bright vitreous luster. Burnt shale MAs are distinguishable by their pinkish-gray color, dull vitreous luster, and platy fracture.

Ferruginous slag MAs are strongly magnetic with agranular vesicular texture. They are distinguishable from coal cinders by a jagged hacky fracture and a metallic luster (Fig. 5.5). Corroded iron nail MAs are distinguished by a metallic luster, and the association of ferrite, goethite and ferrihydrite. Fly ash is comprised partly of dark colored, magnetic, opaque microspheres with metallic luster. Drywall MAs are white, distinguished by very fine pores, and contained plastic fibers. In contrast, bone is characterized by very coarse cavities and pores.

5.3.2 Specific Gravity

The specific gravity (SG) of an artifact is expected to be generally related to the SG of its principal mineral constituent, or constituents (Table 5.3). However, Howard and Orlicki (2015) found that artifacts had SGs lower than those of their principal constituents, presumably as a result of an indeterminant amount of microporosity. Metalliferous slag probably had a SG greater than that of the ferromagnesian minerals ($\sim$3.3) of which it is mainly comprised because it also contained denser phases such as hematite and magnetite with SGs of $\sim$5.2. The SGs of corroded nails were likely variable because of fluctuations in the amounts of remnant ferrite (SG = 4.9) comprising nail cores. Glass slag, resulting from iron smelting, was much denser than the typical soda-lime window and bottle glass, presumably due a high Fe content, or to inclusions of iron oxide or ferromagnesian silicates. Calcareous artifacts such as concrete and mortar were generally denser than siliceous ceramics, probably because of the presence of quartzose aggregate. Concrete may have had a greater value of SG because it contained gravel-sized aggregate, whereas mortar was comprised of sand-sized aggregate. Overall, the SGs of artifacts could be generally categorized as ferruginous > calcareous $\geq$ siliceous > carbonaceous.

Terracotta had a higher SG than the other ceramics studied possibly because it was fired at a lower temperature, and thus contained relatively intact clay minerals. Phyllosilicates have higher SGs ($\sim$2.6) than glass and other low-density high-temperature phases such as cristobalite (SG = 2.3). Carbonaceous artifacts had the lowest SGs attributed to the low density of bitumen ($\sim$1.0). Coal cinders had a much higher SG than other artifacts in the carbonaceous category because they are typically comprised of glass and high density phases such as magnetite (Ward and French 2005). Overall, there was a sort of natural averaging of SG in artifacts comprised of composite materials. The SG of artifacts is expected to contribute to the overall elevated bulk density typical of urban soils. This may be important because previous work suggests that electrical conductivity surveying is sometimes useful for mapping soils on the basis of bulk density (Brevik and Fenton 2004; Islam et al. 2014a, b).

Table 5.3 Chemical and geophysical characteristics of reference artifacts potentially found in urban soils

Sample	Specific gravity x	S	CV (%)	Abrasion pH x	S	CV (%)	Electrical conductivity (μS cm^{-1}) x	S	CV (%)	Resistivity (ρ) (Ω m) ρ	Corrosivity Index	Mass magnetic susceptibility (10^{-8} m^3 g^{-1}) x	S	CV (%)
Carbonaceous														
Wood (Peat)	<1	–	–	3.42	0.02	0.6	557.5	6.4	1.1	17.9	Strong	−3.7	−1.6	43.7
Charcoal	<1	–	–	7.17	0.04	0.6	350.2	28.0	8.0	28.6	Moderate	−4.36	2.14	49.1
Coal	1.19	0.91	7.6	4.53	0.44	9.7	569.0	9.9	1.7	17.6	Strong	−12.0	−0.3	2.3
Coked coal	1.03	0.04	3.9	7.96	0.18	2.2	401.2	3.0	0.7	24.9	Moderate	63.5	1.9	3.0
Coal cinders	2.19	0.08	3.8	8.46	0.35	4.1	149.8	3.2	2.1	66.8	Weak	568.9	40.7	7.1
Coal ash	<1	–	–	6.75	0.15	2.2	478.5	9.2	1.9	20.9	Moderate	880.5	21.5	2.4
Asphaltic concrete	2.31	0.18	7.9	7.52	0.22	2.9	436.3	20.6	4.7	22.9	Moderate	226.2	–	–
Calcareous														
Concrete	2.77	0.14	5.0	12.15	0.05	0.4	10,341.7	426.6	4.1	1.0	Very strong	42.7	14.6	34.1
Pink Mortar	2.32	0.10	4.5	8.80	0.02	0.2	5210.0	1160	22.3	1.92	Very strang	78.0	11.9	15.2
White mortar	2.25	0.06	2.8	11.62	0.01	0.1	2106.7	42.4	13.8	4.7	Very strong	29.1	4.1	14.2
Cinder block	1.64	0.12	7.4	9.72	0.05	0.5	946.0	69.3	7.3	10.6	Strong	356.9	2.0	0.6
Lime brick	2.42	0.16	6.7	8.92	0.7	0.1	1046.0	12.7	1.2	9.6	Strong	54.4	1.4	2.5
Siliceous														
Window glass	2.43	0.05	2.2	9.87	0.17	1.7	262.5	61.5	23.4	38.1	Moderate	−1.0	0	0
Bottle glass	2.50	0.03	1.2	10.60	0.12	1.2	660.6	225.3	34.1	15.1	Strong	3.5	1.7	49.5
Glass slag	2.73	0.04	1.3	9.80	0.11	1.1	103.4	18.2	17.6	96.7	Weak	186.4	2.0	1.1

(continued)

5.3 Microartifacts

Table 5.3 (continued)

Sample	Specific gravity			Abrasion pH			Electrical conductivity (μS cm^{-1})			Resistivity (ρ) (Ω m)		Mass magnetic susceptibility (10^{-8} m^3 g^{-1})		
	X	S	CV (%)	X	S	CV (%)	X	S	CV (%)	ρ	Corrosivity Index	X	S	CV (%)
Red brick	2.37	0.05	2.0	9.11	0.01	0.1	439.2	41.4	9.4	22.8	Moderate	121.0	1.0	0.8
Orange brick	2.25	0.07	3.1	8.31	0.15	1.8	5082.3	170.4	3.4	2.0	Strong	256.8	4.2	1.6
Yellow brick	2.58	0.07	2.6	8.80	0.05	0.5	183.8	10.2	5.6	2.0	Very strong	87.3	1.0	1.1
Terracotta	2.42	0.04	1.8	10.02	0.58	5.8	557.5	71.0	12.7	17.9	Strong	24.7	1.0	4.1
Glazed ceramic pipe	2.07	0.13	6.1	5.51	0.11	1.9	177.4	14.2	8.0	56.4	Weak	89.4	–	–
Ferruginous														
Metalliferous slag	3.66	0.18	4.8	11.50	0.04	0.3	1747.5	296.3	16.9	5.7	Very strong	2010.9	186.7	9.3
Corroded iron	3.44	1.6	45.5	7.69	0.08	0.1	1474.1	203.1	13.8	21.1	Moderate	2786.2	2537	91.0
Miscellaneous														
Bone	1.66	0.02	1.1	7.93	0.43	5.4	615.5	34.6	5.6	16.2	Strong	1.0	0.1	1.9
Drywall	1.07	0.01	1.3	7.13	0.74	10.3	2335.0	35.4	1.5	4.3	Very strong	−8.1	1.9	23.7

X mean; S standard deviation; CV coefficient of variation

5.3.3 Abrasion pH

There is generally a correlation between the measured abrasion pH values of artifacts (Table 5.3) and their principal mineral constituents, e.g., drywall (gypsum), bone (apatite), metalliferous slag (merwinite), corroded nails (goethite), etc. Measured pHs are generally in the order of calcareous > siliceous > ferruginous > carbonaceous. The pHs of calcareous AMPs are well above the abrasion pH (8.2) of calcite, but this can be explained by the very high abrasion pH (~ 12.4) of lime and portlandite. The pH of fresh mortar and concrete is typically ~ 12.0, and drops to 8.0–9.0 after aging and setting via carbonation (Wilimzig and Bock 1996). The very high pH of metalliferous steel-making slag is attributed to the use of limestone or dolostone as a fluxing agent during the iron smelting process. It is well established that Na K, SO_4 and especially Ca become loosely attached to the surface of fly ash particles during the combustion process, and are readily solubilized. The pH of a fly ash-water system is thought to be controlled by the amount of Ca in the ash (Akinyemi et al. 2011; Izquierdo and Querol 2012).

Abrasion pH was attributed previously to the hydrolysis of cations released from the abraded surfaces of minerals (Stevens and Carron 1948; Grant 1969). The high pH values observed for glass and some ceramic AMPs can be explained by hydrolysis of basic cations (particularly Na^+ and K^+) because the surfaces of particles of artifacts are no doubt severely abraded during crushing. For example, hydrolysis of Na released from the abraded surface of a phyllosilicate or glass may occur as follows:

$$\text{Surface} - Na^+ + H_2O \rightarrow \text{Surface} - H^+ + Na^+ + OH^- \quad (5.3)$$

Calcareous artifacts containing unhydrated lime can produce alkalinity by hydrolysis:

$$CaO + H_2O \rightarrow Ca^{2+} + 2OH^{-1} \quad (5.4)$$

Alternatively, calcite or portlandite (limey cement, carbonate lithics as aggregate, etc.) may hydrolyze and generate alkalinity via bicarbonate production:

$$CaCO_3 + H_2CO_3 \rightarrow Ca^{2+} + 2HCO_3 \quad (5.5)$$

$$Ca(OH)_2 + H_2CO_3 \rightarrow Ca^{2+} + HCO_3 + H_2O + OH^- \quad (5.6)$$

$$HCO_3 + H_2O \rightarrow H_2CO_3 + OH^- \quad (5.7)$$

Some ceramic artifacts have a slightly acidic pH possibly because of hydrolysis of Al^{3+} (or Fe^{3+}) exposed at the broken edges of phyllosilicates or other phases:

5.3 Microartifacts

$$\text{Surface} - \text{Al}^{3+} + \text{H}_2\text{O} \rightarrow \text{Surface} - \text{Al(OH)}^{2+} + \text{H}^+ \quad (5.8)$$

Wood, charcoal and coal may be acidic through dissociation of carboxyl or other organic functional groups:

$$\text{R} - \text{COOH} \rightarrow \text{R} - \text{COO}^- + \text{H}^+ \quad (5.9)$$

5.3.4 Trace Metal Content

The artifacts analyzed from urban anthropogenic soils in Detroit, Michigan (Table 5.4) were found to contain a wide variety of trace metals and other elements, similar to results obtained elsewhere (e.g., Shaw et al. 2010). Brick and terracotta contain concentrations of Pb, As, Zn, Ni, Cd, Mn, Cu and Ti which were elevated with respect to background levels commonly found in soils. Mortar and concrete contained a similar assemblage of trace elements, but Cr was below detection limits. They also contained elevated levels of S which is a well known component of Portland cement. Iron slag contained relatively high concentrations of Ni and Cr along with significant amounts of Ti and S. Slag contained especially high levels of Mn, which is an essential additive in the steel-making process. Coal contained some Pb, As, Ti and S, but many of the other elements were below detection limits. In contrast, coal combustion products (ash, cinders, cinder block) were distinguished by the presence of Hg, and contained elevated levels of all elements except Cd. Coal ash was especially rich in As and Ti. The presence of heavy metals in artifacts

Table 5.4 Total elemental concentrations in various artifacts measured by X-ray fluorescence spectroscopy

Sample type	Element (mg kg^{-1})								
	Pb	As	Hg	Zn	Ni	Cd	Mn	Cr	Ti	S
Org. Brk	28	14	bdl	75	40	bdl	650	59	3366	bdl
Red brick	45	11	bdl	87	70	8	496	59	4285	bdl
Mortar	8	6	bdl	23	48	9	315	bdl	671	4190
Concrete	8	4	3	51	50	10	333	bdl	666	8793
Iron-slag	25	2	35	96	580	13	64,514	1103	1850	1107
Coal ash	90	35	18	115	129	bdl	137	208	12,098	839
C. cinders	25	7	4	36	123	7	804	128	9969	439
Coal	6	3	8	bdl	bdl	bdl	bdl	bdl	260	5976
Cind. blk	45	14	bdl	103	74	9	563	67	4510	1539
Terracotta	34	11	bdl	109	82	7	682	111	6757	bdl
X	31	13	7	70	120	6	6849	174	4443	2288
S	25	10	11	40	166	5	20,263	333	4061	3032

X mean; S standard deviation; *bdl* below detection limit

is important partly because of their potential release to the soil environment as a result of artifact weathering. It is also important when total metal concentrations are determined, e.g., by X-ray fluorescence spectroscopy, because it can result in a greatly overestimated assessment of environmental risk. Many studies have shown a strong correlation between soil magnetic susceptibility and toxic metal content, thus facilitating the geospatial analysis of contaminated soils (Vodyanitskii and Shoba 2015).

5.3.5 Electrical Conductivity

Measured values of electrical conductivity (EC) range from 103 to 10, 342 $\mu S\ cm^{-1}$ for a wide variety of artifact types (Table 5.3). Corresponding values of electrical resistivity (ER) ranged from 1.0 to 96.7 Ωm. The highest ECs are for calcareous waste building materials and metalliferous slag. Bone and carbonaceous AMPs are intermediate, and glassy AMPs have the lowest ECs. Based on ERs, and the corrosivity index used for metals buried in soils (Elias 2000), crushed artifactual materials comprised of waste building materials are strongly corrosive. It is well established that silicates, non-metallic non-silicates, and glass are poor electrical conductors. However, if such materials are in contact with water, an electrical current can propagate by ionic conduction. Thus, ECs and ERs of soil and rock are generally dependent on the electrolytic characteristics of associated pore fluids (Rhoades et al. 1989; Telford et al. 1990; Corwin and Lesch 2005). Ionic conduction may be affected by electrically charged particle surfaces associated with phyllosilicates (Kriaa et al. 2014) and calcite (Wu et al. 2010) because the diffuse double layer has a higher conductivity than the bulk solution (Telford et al. 1990; Wightman et al. 2003). EC can also occur by electron transfer resulting from hydrogen and transition metal impurities. EC is generally thought to be proportional to ferric iron content (Schaefer 2010; Karato and Wang 2013).

Orange brick artifacts are common at demolition sites in Detroit, Michigan where buildings from the 19th century were razed (Howard and Orlicki 2015). The very high EC of these artifacts (Table 5.3) is attributed to primitive technology, and firing at <900 °C, which left clay minerals relatively intact. Phyllosilicates decompose into glass above ~900 °C (Cultrone et al. 2004, 2005; Reedy 2008). Hence, the lower EC of red brick dating from the 20th century is ascribed to higher glass content resulting from improved technology and firing at a temperature >900 °C. The pink mortar studied, collected from a derelict masonry building built before ~1930, was probably made by adding brick dust. Hence, it had an EC nearly as high as the orange brick. The elevated EC of drywall is attributed to its high microporosity, and enhanced electrolytic conduction caused by solution of Ca and SO_4 released from abraded surfaces of gypsum particles. Bone had a much lower EC, presumably due to the lower solubility of apatite. The relatively high ECs of metalliferous slag, red brick and cinder block may be the result of electron transfer reactions involving elemental iron and Fe-oxides. Similarly, electron transfer reactions involving H impurities may account for the observed ECs of carbonaceous AMPs. Glass is an insulator, hence the lowest ECs were associated with glassy AMPs.

5.3.6 Magnetic Susceptibility

Table 5.3 shows that measured values of magnetic susceptibility (MS) are highest for corroded iron nail, metalliferous slag, coal cinder, coal ash, and cinder block artifacts. All of these artifact types responded to a hand magnet indicating the presence of ferrimagnetic material. The high MS of corroded nails was attributed to the presence of variable amounts of uncorroded remnants of a ferrite (elemental Fe) nail core. The MS of metalliferous slag, which is nearly as high, can be ascribed to the fact that hematite can be converted into magnetite by oxidation at 900–1000 °C (Hu et al. 2006), or rapidly (<3 min) at ~600 °C by reduction with only a small amount H_2 (Matthews 1976; Kontny and Dietl 2002; Wagner et al. 2006; Bhargava et al. 2009). Similarly, the elevated MS of cinder block is attributed to the presence of magnetite-bearing blast furnace slag as aggregate. Coal often contains pyrite that decomposes into hematite or magnetite at temperatures of 200–1000 °C (Waanders et al. 2003; Hu et al. 2006). Hence, the high values of MS obtained for coal cinders and ash is explained by magnetite formed as a by-product of coal combustion. It is well established that the degree of magnetization in response to an applied magnetic field is generally a function of the amount of magnetite, maghemite, elemental iron and heavy metals present in an earth material (Oldfield 1991; Verosub and Roberts 1995; Schmidt et al. 2005; Magiera et al. 2006).

References

Abdullah MI, Atherton MP (1964) The thermometric significance of magnetite in low grade metamorphic rocks. Am J Sci 262:904–917
Ahmad S, Iqbal Y, Ghani F (2008) Phase and microstructure of brick-clay soil and fired clay-bricks from some areas in Peshawar Pakistan. J Pak Mater Soc 2:33–39
Akinyemi SA, Akinlua A, Gitari WM, Akinyeye RO, Petrik LF (2011) The leachability of major elements at different stages of weathering in dry disposed fly ash. Coal Comb Gasif Prod 3:28–52
Asami K, Kikuchi M (2002) Characterization of rust layers on weathering steels air-exposed for a long period. Mater Trans 43:2818–2825
Barrow CJ (2012) Biochar: potential for countering land degradation and for improving agriculture. Appl Geog 34:21–28
Bayless ER, Bullen TD, Fitzpatrick JA (2004) Use of $^{87}Sr/^{86}Sr$ and $\delta^{11}B$ to identify slag-affected sediment in southern Lake Michigam. Environ Sci Technol 38:1330–1337
Berna F, Matthews A, Weiner S (2004) Solubilities of bone mineral from archaeological sites: the recrystallization window. J Arch Sci 31:867–882

Bhargava SK, Garg A, Subsinghe ND (2009) In situ high-temperature phase transformation studies on pyrite. Fuel 88:988–993

Brevik EC, Fenton TE (2004) The effect of changes in bulk density on soil electrical conductivity as measured with the Geonics® EM-38. Soil Surv Horiz 45(3):96–102

Brodowski S, Amelung W, Haumaier L, Abetz C, Zech W (2005) Morphological and chemical properties of black carbon in physical soil fractions as revealed by scanning electron microscopy and energy dispersive X-ray spectroscopy. Geoderma 128:116–129

Brown JL (2013) Rocky road: the story of asphalt pavement. Civil Eng 83:40–43

Carlson CL, Adriano DC (1993) Environmental impacts of coal combustion residues. J Environ Qual 22:227–247

Chang S, Berner RA (1999) Coal weathering and the geochemical carbon cycle. Geochim Cosmo Acta 63:3301–3310

Choudury N, Mohanty D, Boral P, Kumar S, Hazra SK (2008) Microscopic evaluation of coal and coke for metallurgical usage. Curr Sci 94:74–81

Cornelissen G, Gustafsson O, Bucheli TD, Jonker MTO, Koelmans AA, van Noort PCM (2005) Extensive sorption of organic compounds to carbon black, coal, and kerogen in sediments and soils: mechanisms and consequences for distribution, bioaccumulation, and biodegradation. Environ Sci Technol 39:6881–6895

Corwin DL, Lesch SM (2005) Apparent soil electrical conductivity measurements in agriculture. Comput Electron Agric 46:11–43

Cultrone G, Sebastian E, Elert K, de la Torre MJ, Cazalla O, Rodriguez-Navarro C (2004) Influence of mineralogy and firing temperature on the porosity of bricks. J Eur Ceram Soc 24:547–564

Cultrone G, Sebastian E, de la Torre MJ (2005) Mineralogical and physical behavior of solid bricks with additives. Constr Build Mater 19:39–48

Diamant RME (1970) The chemistry of building materials. Business Books, London

Dunnel RC, Stein JK (1989) Theoretical issues in the interpretation of microartifacts. Geoarchaeology 4:31–42

Duval DJ, Risbud SH, Shackelford JF (2008) Mullite. In: Shackelford JF, Doremus RH (eds) Ceramic and glass materials: 27 structure, properties and processing. Springer, Berlin

Elias V (2000) Corrosion/degradation of soil reinforcements for mechanically stabilized earth walls and reinforced soil slopes. U.S. Department of Transportation. Federal Highway Administration Pub. FHWA-NHI-00-044: 94 p

Fisher GL, Chang DPY, Brummer M (1976) Fly ash collected from electrostatic precipitators: microcrystalline structures and the mystery of the spheres. Science 192:553–555

Fisher GL, Prentice BA, Silberman D, Ondov JM, Biermann AH, Ragaini RC, McFarland AR (1978) Physical and morphological studies of size-classified coal fly ash. Environ Sci Technol 12:447–451

Forbes MS, Raison RJ, Skjemstad JO (2006) Formation, transformation and transport of black carbon (charcoal) in terrestrial and aquatic systems. Sci Total Environ 370:190–206

Fredericci C, Zanotto ED, Ziemath EC (2000) Crystallization mechanism and properties of a blast furnace slag glass. J Non-Cryst Solids 273, 64–75

Fromm J (2013) Xylem development of trees: from cambial divisions to mature wood cells. In: Fromm J (ed) Cellular aspects of wood formation. Plant cell monographs, vol 20. Springer, Berlin, pp 3–40

Grant WH (1969) Abrasion pH, an index of chemical weathering. Clays Clay Miner 17:151–155

Gray RJ (1991) Some petrographic applications to coal, coke and carbons. Org Geochem 17:535–555

Heitz E (1996) Electrochemical and chemical mechanisms. In: Heitz E, Flemming HC, Sand W (eds) Microbially influenced corrosion of materials. Springer, Berlin, pp 27–38

Hofrichter M, Fakoussa RM (2001) Microbial degradation of coal. In: Hofrichter M, Steinbuchel A (eds) Biopolymers v. 1: Lignin, humic substances and coal. Wiley VCH, pp 393–429

References

Howard JL, Olszewska D (2011) Pedogenesis, geochemical forms of heavy metals, and artifact weathering in an urban soil chronosequence, Detroit, Michigan. Environ Pollut 159:754–761

Howard JL, Orlicki KM (2015) Effects of anthropogenic particles on the chemical and geophysical properties of urban soils, Detroit, Michigan. Soil Sci 180:154–166

Howard JL, Orlicki KM (2016) Composition, micromorphology and distribution of microartifacts in anthropogenic soils, Detroit, Michigan USA. Catena 138:38–51

Howard JL, Dubay BR, Daniels WL (2013) Artifact weathering, anthropogenic microparticles, and lead contamination in urban soils at former demolition sites, Detroit. Michigan Environ Pollut 179:1–12

Howard JL, Ryzewski K, Dubay BR, Killion TK (2015) Artifact preservation and post-depositional site-formation processes in an urban setting: a geoarchaeological study of a 19th century neighborhood in Detroit. Michigan J Archaeol Sci 53:178–189

Howard JL, Orlicki KM, LeTarte SM (2016) Evaluation of geophysical methods for mapping soils in urbanized terrain, Detroit, Michigan USA. Catena 143: 145–158

Hu G, Dam-Johansen K, Wedel S, Hansen JP (2006) Decomposition and oxidation of pyrite. Prog Energy Combust Sci 32:295–314

Hunt CB (1953) Dating of mining camps using tin cans and bottles. Geotimes 3:8–10

ICCP (International Committee for Coal and Organic Petrology) (1998) The new vitrinite classification (ICCP System 1994). Fuel 77:349–358

Islam MM, Meerschman E, Saey T, De Smedt P, Van De Vijver E, Delefortrie S, Van Meirvenne M (2014a) Characterizing compaction variability with an electromagnetic induction sensor in a puddled paddy rice field. Soil Sci Soc Am J 78:579–588

Islam MM, Saey T, Smedt P, Van De Vijver E, Delefortrie S, Van Meirvenne M (2014b) Modeling within field variation of the compaction layer in a paddy rice field using a proximal soil sensing system. Soil Use Manag 30:99–108

IUSS Working Group (2015) World Reference Base for Soil Resources 2014 (update 2015), International soil classification system for naming soils and creating legends for soil maps. World Soil Resources Reports No. 106. FAO, Rome, Italy

Izquierdo M, Querol X (2012) Leaching behavior of elements from coal combustion fly ash: an overview. Int. Jour. Coal Geol. 94:54–66

Karato SI, Wang D (2013) Electrical conductivity of minerals and rocks. In: Karato SI (ed) Physics and chemistry of the deep Earth. Wiley, New York, pp 145–182

Kosmatka SH, Kerkoff B, Panerese WC (2002) Design and control of concrete mixtures, 14th edn. Portland Cement Association, Skokie, IL, p 358

Kontny A, Dietl C (2002) Relationships between contact metamorphism and magnetite formation and destruction in a pluton's aureole, White-Inyo Range, eastern California. Geol Soc Am Bull 114:1438–1451

Kriaa A, Hajji M, Jamoussi F, Hamzaoui, AH (2014) Electrical conductivity of 1:1 and 2:1 clay minerals. Surf. Eng. Appl. Electrochem. 50: 84–94

Krull ES, Skjemstad JO (2002) $\delta^{13}C$ and $\delta^{15}N$ profiles in ^{14}C dated Oxisol and Vertisols as a function of soil chemistry and mineralogy. Geoderma 1890:1–29

Lane DS (2004) Petrographic methods of examining hardened concrete: a petrographic manual. U. S. Department of Transportation Pub. FHWA-HRT-04-150, 324 p

Lanteigne S, Schindler M, McDonald AM, Skeries K, Abdu Y, Mantha NM, Murayama M, Hawthrone FC, Hochella MF Jr (2012) Mineralogy and weathering of smelter-derived spherical particles in soils: Implications for the mobility of Ni and Cu in the surficial environment. Water Air Soil Pollut 223:3619–3641

Livingston RA, Stutzman PE, Schumann I (1998) Quantitative X-ray diffraction analysis of handmolded brick. In: Baer NS, Fritz S, Livingston RA (eds) Conservation of historic brick structures. Donhead Publishing. Ltd., Shafesbury, UK, pp 105–116

Lukasik A, Szuszkiewicz M, Magiera T (2015) Impact of artifacts on topsoil magnetic susceptibility enhancement in urban parks of the Upper Silesia conurbation datasets. J Soils Seds 15:1836–1846

Lu SG, Wang HY, Guo JL (2011) Magnetic enhancement of urban roadside soils as a proxy of degree of pollution by traffic-related activities. Environ Earth Sci 64:359–371

Magiera T, Strzyszcz Z, Kapicka A, Petrovsky E (2006) Discrimination of lithogenic and anthropogenic influences on topsoil magnetic susceptibility in central Europe. Geoderma 130:299–311

Martinez M, Escobar M (1995) Effect of coal weathering on some geochemical parameters. Org Geochem 23:253–261

Matthews A (1976) Magnetite formation by the reduction of hematite with iron under hydrothermal conditions. Am Mineral 61:927–932

Minyuk PS, Tyukova EE, Subbotnikova TV, Kazansky AY, Fedotov AP (2013) Thermal magnetic susceptibility data on natural iron sulfides of northeastern Russia. Russian Geol Geophy 54:464–474

Mukherjee S (2011) Applied mineralogy—applications in industry and environment. Springer, Dordrecht 575 p

Neff D, Dillman P, Bellot-Gurlet L, Beranger G (2005) Corrosion of iron archaeological artefacts in soil: characterization of the corrosion system. Corros Sci 47:515–535

Nguyen BT, Lehmann J, Kin yangi J, Smernik R, Riha SJ, Engelhard MH (2008) Long-term black carbon dynamics in cultivated soil. Biogeochemistry 89:295–308

Oldfield F (1991) Environmental magnetism—a personal perspective. Quat Sci Rev 10:73–85

Petrakis L, Grady DW (1980) Coal analysis, characterization and petrography. J Chem Educ 57:689–694

Piatak NM, Seal RR II (2012) Mineralogy and environmental geochemistry of historic iron slag, Hopewell Furnace National Historic Site, Pennsylvania. USA Appl Geochem 27:623–643

Proctor DM, Fehling KA, Shay EC, Wittenborn JL, Green JJ, Avent C, Bigham RD, Connolly M, Lee B, Shepker TO, Zak MA (2000) Physical and chemical characteristics of blast furnace, basic oxygen furnace, and electric arc furnace steel industry slags. Environ Sci Technol 34:1576–1582

Rapp G (2009) Archaeomineralogy. Springer, Dordrecht 348 p

Reedy CL (2008) Thin-section petrography of stone and ceramic cultural materials. Archetype Publications Ltd., London 256 p

Rhoades JD, Manteghi NA, Shouse PJ, Alves WJ (1989) Soil electrical conductivity and soil salinity: New formulations and calibrations. Soil Sci. Soc. Am. Jour. 53: 433–439

Roberts FL, Kandhal PS, Brown RE, Lee DY, Kennedy TW, Thomas W (1994) Hot mix asphalt materials, mixture design, and construction, NAPA Research and Education Foundation

Rutherford DW, Wershaw RL, Cox LG (2005) Changes in composition and porosity occurring during the thermal degradation of wood and wood components. USGS Special Investment Report 2004–5292, 79 p

Schaefer MV (2010) Spectroscopic evidence for interfacial Fe(II)-Fe(III) electron transfer in clay minerals. M.S. thesis, Department of Civil and Environmental Engineering, University of Iowa, Ames, IA

Schiffer MB (1987) Formation processes of the archaeological record. University of New Mexico Press, Albuquerque, NM 428 p

Schmidt A, Yarnold R, Hill M, Ashmore M (2005) Magnetic susceptibility as proxy for heavy metal pollution: a site study. J Geochem Explor 85:109–117

Schwertmann U, Fisher WR (1973) Natural "amorphous" ferric hydroxide. Geoderma 10:237–247

Shaw RK, Wilson MA, Reinhardt L, Isleib J (2010) Geochemistry of artifactual coarse fragment types from selected New York City soils. 19th World Cong. Soil Sci., Bribane, Australia, 25–27

Sherwood SC (2001) Microartifacts. In: Goldberg P, Holliday VT, Ferring CR (eds) Earth sciences in archaeology, Kluwer Academic, New York, pp 327–351

Singh V, Pande PC, Jain DK (2010) Text book of botany: structure development and reproduction in Angiosperms. Rastogi Publications, 220 p

Smith RD, Campbell JA, Nielson KK (1979) Concentration dependence upon particle size of volatized trace elements in fly ash. Environ Sci Technol 13:553–558

References

Soil Survey Staff (2014) Keys to soil taxonomy, 12th edn. USDA-NRCS, U.S. Government Print Office, Washington, DC

Stevens RE, Carron MK (1948) Simple field test for distinguishing minerals by abrasion pH. Am Minerl 33:31–49

Stoltman JB (2001) The role of petrography in the study of archaeological ceramics. In: Goldberg P, Holliday VT, Ferring CR (eds) Earth sciences and archaeology. Plenum Publishers, New York, pp 297–326

Suarez-Ruiz I (2012) Organic petrology: An overview. In: Al-Juboury A (ed) Petrology—new perspectives and applications. Shanghai, Intech Publishers, pp 199–224

Telford WM, Geldart LP, Sheriff RE (1990) Applied geophysics, 2nd edn. Cambridge University Press, pp 283–292

Van Oss HG (2005) Background facts and issues concerning cement and cement data. U.S. Geological Survey Open File Report 2005–1152, 44 p

Verosub KL, Roberts AP (1995) Environmental magnetism: past, present and future. J Geophys Res 100(B2):2175–2192

Vodyanitskii YN, Shoba SA (2015) Magnetic susceptibility as an indicator of heavy metal contamination in urban soils (Review). Moscow Univ Soil Sci Bull 70:13–20

Wagner D, Devisme O, Patisson F, Ablitzer D (2006) A laboratory study of the reduction of iron oxides by hydrogen. Proc Sohn Int Symp 2:111–120

Waanders FB, Vinken E, Mans A, Mulaba-Bafubiandi AF (2003) Iron minerals in coal, weathered coal and coal ash—SEM and Mossbauer results. Hyperfine Interact 148(149):21–29

Ward CR (2002) Analysis and significance of mineral matter in coal seams. Int J Coal Geol 50:135–168

Ward CR, French D (2005) Relation between coal and fly ash mineralogy, based on quantitative X-ray diffraction methods. Proc World Coal Ash Conf, Lexington, KY 14 p

Wightman WE, Jalinoos F, Sirles P, Hanna K (2003) Application of geophysical methods to highway related problems. Federal Highway Administration, Central Federal Lands Highway Division, Lakewood, CO, Publication No. FHWA-IF-04-02

Wilimzig M, Bock E (1996) Attack of mortar by bacteria and fungi. In: Heitz E, Flemming HC, Sand W (eds) Microbially-influenced corrosion of materials. Springer, Berlin, pp 311–323

Wu Y, Hubbard S, Williams KH, Ajo-Franklin J (2010) On the complex conductivity signatures of calcite precipitation. J Geophys Res 115, G00G04

Yang T, Liu Q, Li H, Zeng Q, Chan L (2010) Anthropogenic magnetic particles and heavy metals in the road dust: Magnetic identification and its implications. Atmos Environ 44:1175–1185

Yang T, Liu Q, Zeng Q, Chan L (2012) Relationship between magnetic properties and heavy metals of urban soils with different soil types and environmental settings: implications for magnetic mapping. Environ Earth Sci 66:409–420

Yildirim IZ, Prezzi M (2011) Chemical, mineralogical, and morphological properties of steel slag. Adv Civil Eng Article 463638, 13 p

Chapter 6
Classification of Anthropogenic Soils

In U.S. Soil Taxonomy, anthropogenic soils are classified into two basic categories depending on type of soil parent material: (1) human-altered material (HAM), and (2) human-transported material (HTM). Generally speaking, a surface layer of HTM or HAM ≥ 50 cm thick must be present in order for a pedon to be classified using the anthropogenic soil profile. If so, the pedon is classified according to soil order, and then further as Anthropic, Anthrodensic, or Anthroportic subgroups if developed in HTM, or as Anthraltic subgroups if developed in HAM. Most anthropogenic soils are classified as Entisols, although they sometimes contain a cambic-like B horizon. Some urban and mine soils contain cambic B horizons and are classified as Inceptisols. Some anthropogenic soils on ancient burial mounds have well developed cambic and argillic horizons, and are classified as Inceptisols, Alfisols, and occasionally Ultisols. Anthropogenic soils are generally classified as Anthrosols and Technosols in the World Reference Base. An Anthrosol is a natural soil that has been drastically modified by long-term human additions of organic and/or inorganic materials. Anthrosols are found at archaeological sites of human habitation (kitchen middens, grave soils, etc.), and in areas where agricultural activities have been practiced for periods of many centuries (garden soils, Amazonian Dark Earth, paddy soils, etc.). A Technosol is characterized by the presence of technogenic artifacts, or by parent materials that are present at the Earth's surface only because of human action (e.g., mine spoils). Technosols include mine soils, burial mound soils, urban soils, and land surfaces sealed by artificial materials (e.g., pavement).

6.1 Introduction

According to Simonson (1962), the earliest known classification of soils was developed in China during the Yao dynasty (2357–2261 B.C.) about 4000 years ago. Soils were classified based on agricultural productivity into nine different types, which determined the value and amount of taxation of land holdings by individuals and families. According to medieval scholars, the ancient Greeks and Romans classified soils for similar reasons. Soils continued to be studied and classified based on agricultural productivity, but during the late 19th century classifications based on geology began to appear. Dokuchaiev had begun classifying soils in Russia based on their natural characteristics (e.g., soil profile morphology) during the late 1800s, but this approach did not reach the rest of Europe and North America until the early 20th century. Most modern systems of soil classification have been developed only within the last 50 years or so (e.g., Soil Survey Staff, 1975), and most efforts to classify anthropogenic soils have materialized only within the last 10 or 15 years.

Modern taxonomic classifications are based on the natural characteristics of soil. Most classification systems have multiple categories (levels) of classification, which in turn are classified into separate taxa (classes), based on definitions of differentiating characteristics. A **category** is a set of classes defined by the same level of generalization or abstraction. A **class** is defined as a group of individuals, or of other classes, with similar properties, which can be distinguished from other classes by these properties. These properties are referred to as **differentiating characteristics**, which serve as the basis for soil classification. The presence or absence of certain kinds of surface and subsurface horizons are typically used as differentiating characteristics to separate categories in soil classification, whereas taxa are distinguished by more specific properties such as color, texture, structure, consistence, mineralogy, temperature, cation exchange capacity, etc. (Fanning and Fanning 1989). Differentiating characteristics include properties such as texture, which are useful because a number of other properties are correlated with them. For example, water-holding capacity, hydraulic conductivity, organic matter content and other characteristics are correlated with texture. These other properties which correlate with a differentiating characteristic are called **accessory characteristics**. Thus, a good differentiating characteristic has a large number of accessory characteristics. Properties not correlated with a differentiating characteristic are called **accidental characteristics**. For example, slope varies independently of texture, and is thus an accidental characteristic of a class defined on the basis of texture (Buol et al. 2011).

Anthropogenic soils in urban and mine-related geocultural settings are generally characterized by soil profiles with evidence of very weak to weak pedogenic development. This is because such sites are generally of modern origin and so there has been a limited amount time for pedogenic processes to operate, or because pre-existing pedogenic features were obliterated by anthroturbation. In contrast, at locations where agricultural activities persisted over very long periods of time (perhaps many centuries), anthropogenic soil profiles are characterized by moderately to well developed pedogenic horizons. There are also very well developed profiles known to be present at certain archaeological sites where anthropogenic soils have developed in deposits of fill on land surfaces (e.g., burial mounds) that have been stable for many centuries or even millennia.

Excellent overviews of soil classification systems from many different countries are found in Eswaran et al. (2003), Rossiter (2007), Lehmann and Stahr (2007),

6.1 Introduction

Krasilnikov et al. (2010), Buol et al. (2011), and Capra et al. (2015). There are also excellent discussions of soil classifications and anthropogenic soils in Australia (Isbell, 2016), China (Gong et al. 2003; Shi et al. 2006; Gong and Zhang 2007), Germany (Burghardt 1994; Meuser and Blume 2001), Poland (Charzynski et al. 2013), and Russian (Alexandrovskaya and Alexandrovskiy 2000; Prokofyeva et al. 2011; Lebedeva and Gerasimova 2012). Other possible articles of interest are those of Bryant and Galbraith (2003), Dazzi et al. (2009), Puskas and Farsang (2009), Gray et al. (2011) and Sarmast et al. (2016).

6.2 U.S. System of Soil Taxonomy

In U.S. Soil Taxonomy (Soil Survey Staff 2014), anthropogenic soils are classified on the basis of two key characteristics: (1) parent material (human-transported material vs. human-altered material), and (2) the presence of diagnostic horizons and other characteristics (Table 6.1). Further special considerations are used in the case of buried soils. A buried soil is a sequence of genetic horizons in a pedon that is covered with a surface mantle of new soil material that is ≥ 50 cm thick, or which is covered by a plaggen epipedon. A surface mantle of new soil material is defined as a layer of natural or human-deposited mineral matter that is unweathered, at least in its lower part. A surface mantle can have an epipedon and/or a cambic horizon, but there remains a layer ≥ 7.5 cm in its lower part that fails the requirements for all diagnostic horizons overlying the buried genetic sequence. Hence, if a surface mantle is ≥ 50 cm thick and the lower part is unaltered, then the soil order is determined based on diagnostic characteristics of the anthropogenic soil profile developed in the surface mantle, whereas if the lower part is altered, then the entire soil profile is used to determine the soil order. If the surface mantle is <50 cm thick, then only the characteristics of the buried natural soil are used for classification (Soil Survey Staff 2011).

Human-altered material (HAM) is parent material for a soil that has undergone some form of intentional human-induced alteration, including deep plowing to break up root-restrictive subsoil layers, excavation (such as for a gravel pit), or surface compaction in order to puddle water (such as for rice paddies). Although highly altered from their natural state, they have not been transported from one location to another. Human-altered materials contain evidence of purposeful human alteration and were either: (1) Tilled to >50 cm to break up a root-restrictive subsoil layer, (2) Located on a destructional anthropogenic landform (such as a gravel pit), or (3) Purposely compacted to puddle water for agricultural purposes. Although it may be designated with the suffix p or d (e.g., Ap and Bdg), there is no specific horizon nomenclature to indicate human-altered material, according to Soil Taxonomy. The problem with this approach is that the anthropogenic nature of the soil is obscured. Hence, elsewhere in this book, anthropogenic horizons formed in HAM are designated informally by the asterisk symbol (*).

Table 6.1 Types of diagnostic horizons and characteristics (HAM, human-altered material; HTM, human-transported material; n/a, not applicable) potentially associated with anthropogenic soil profiles, according to Soil Taxonomy (Soil Survey Staff 2014)

Soil profile characteristic	Horizon designation	Description
Epipedon (Diagnostic surface horizon)		
Ochric	A, ^A	Organic-poor eluvial layer <10–25 cm thick with base saturation <50%
Anthropic	A, ^A	Organic-bearing mineral layer ≥ 25 cm thick formed in HAM or HTM that either overlies mine or dredged spoil directly, or has artifacts, midden material or anthraquic conditions throughout
Plaggen	^A	Organic-rich layer of HTM ≥ 50 cm thick containing artifacts and/or spade marks
Diagnostic subsurface horizon		
Agric	Bh, Bw, Bt	Illuvial horizon ≥ 10 cm thick located directly below a plow layer containing illuvial silt/clay/humus as wormhole linings or thin lenses comprising >5%, by volume
Albic	E	Light-colored eluvial subsoil horizon leached of clay and free iron-oxides to such a degree that the color of the horizon is mostly due to uncoated sand and silt particles
Argillic	Bt	Illuvial horizon ≥ 7.5 cm thick, forms as a result of the translocation of clay from an overlying eluvial layer
Calcic	k	Illuvial horizon ≥ 15 cm thick with a calcium carbonate equivalent $\geq 15\%$ and is 5% higher by weight (absolute) than an underlying horizon, or has $\geq 5\%$, by volume, visible secondary $CaCO_3$
Cambic	w	Mineral horizon of very fine sand, loamy very fine sand, or finer texture, ≥ 15 cm thick with some weak indication of reddening/gleying, structure, removal of carbonate or gypsum, or illuvial clay accumulation
Gypsic	Ay, By	A surface or subsurface horizon >15 cm thick with >1% visible gypsum. Horizon has >1% visible secondary gypsum
Salic	Az, Bz	A saline surface or subsurface horizon >15 cm thick in which salts more soluble than gypsum (e.g., halite) have accumulated.
Other diagnostic characteristics		
Identifiable secondary carbonate	k	Translocated pedogenic $CaCO_3$ that has precipitated within a soil horizon(s)
Lithologic discontinuity	n/a	A significant change in particle size distribution or mineralogy that indicates a difference in lithology or age within a soil profile
n-value	n/a	A measure of the tendency of a soil to fail in a fluid state
Aquic conditions	g	Conditions of periodic or continuous saturation and chemical reduction

(continued)

6.2 U.S. System of Soil Taxonomy

Table 6.1 (continued)

Soil profile characteristic	Horizon designation	Description
Anthric saturation	g	Human-induced aquic conditions in cultivated and flood-irrigated soils
Densic material	d	Horizon with a bulk density (≥ 1.6 to 1.8 g cm^{-3}) such that plant roots cannot enter
Sulfidic material	se	Horizon containing oxidizable sulfide minerals
Sulfuric horizon	j	Horizon ≥ 15 cm thick with a pH ≤ 3.5 caused by sulfuric acid
Diagnostic anthropogenic characteristics		
HAM[a]	A, B, C	Parent material produced by *in situ* human-induced mixing or disturbance
HTM	^A, ^B, ^C	Parent material produced by moving mass from an outside source onto a soil surface at a different location
Artifacts	u	Objects manufactured, modified, or transported from their source by humans
Manufactured layer	M	A root-limiting layer composed of artificial, impermeable materials

[a]Designated as (*) in this book, e.g., *A or *Bw horizon

Human-transported material (HTM) is parent material for a soil which was intentionally moved or imported by human activity from a location away from the site where it is now located (Fig. 6.1). HTM is usually underlain at some depth by a lithologic discontinuity and/or a buried soil horizon from the original soil that existed at the site before deposition of the new anthropogenic material. Horizons composed of human-transported material are designated by the caret symbol (^) as a prefix in combination with master horizons, such as O, A, B, or C. These may be combined with suffix d, j, u, or w (e.g., ^Ap, ^Cj, ^Cdu). Soil material above a manufactured layer (designated by M) is human-transported.

Fig. 6.1 Deposit of human-transported material. Photo by USDA-NRCS

Epipedons are the diagnostic surface horizons of mineral soils which have many properties of, but are not synonymous with, A-horizons. They include the upper part of the soil darkened by organic matter, or the upper eluvial horizons which may become darkened under cultivation, or both, or they may have only lost rock structure. In Soil Taxonomy, rock structure includes fine stratification (less than 5 mm) in unconsolidated sediments, and saprolite derived from consolidated rocks in which the unweathered minerals and pseudomorphs of weathered minerals retain their relative positions to each other in the soil profile (Soil Survey Staff 2014). Epipedons may include part of the illuvial B-horizon if there is enough darkening by organic matter. They may be buried by shallow sediments, which are not part of the epipedon if they retain fine stratification. The properties of epipedons, except for structure, are determined after the upper 18 cm are mixed or, if bedrock is shallower than 18 cm, after the whole soil has been mixed.

An **ochric epipedon** (Table 6.1) is typically the first to form in anthropogenic soils. It is generally a surface layer in which rock structure has been destroyed, which is only slightly or moderately darkened by organic matter, and which shows some minor evidence of eluviation. The ochric epipedon is basically a "catch-all" epipedon for those horizons which fail to meet the requirements of other epipedons. Hence, it may be somewhat similar to any of the other epipedons. An ochric epipedon is generally designated as an A or ^A, often in combination with suffix u or p (e.g., ^Au, Au, Ap). Over time, an anthropogenic ochric epipedon may evolve into an **anthropic epipedon**. This is a relatively thick horizon formed in human-altered or human-transported material that shows evidence of purposeful alteration of soil properties, or of earth-surface features, by human activity (Fig. 6.2). The key feature is that it formed as the result of intentional human alteration (but not simply by the common agricultural practices of plowing and amending the soil with fertilizers). An anthropic epipedon is located on an anthropogenic landform and/or contains artifacts. It may have been compacted to impede drainage (as in paddy cultivation), and some anthropic epipedons meet the color and organic requirements of mollic epipedons. An anthropic epipedon is generally designated as an A or ^A, often with the suffix u or p (e.g., ^Au, Au, Ap). A **plaggen epipedon** is an anomalously thick, dark-colored (typically black to dark grayish brown) human-made mineral surface layer produced by long-term manuring and spading. It typically contains a few artifacts, such as bits of brick and pottery, throughout the epipedon. It is widespread in Europe, but not widely recognized in the United States. A **manufactured layer** is an artificial, root-limiting layer consisting of nearly continuous, human-manufactured materials whose purpose is to form an impervious barrier. It has no cracks, or the spacing of cracks that roots can enter is ≥ 10 cm. The manufactured layer may be composed of asphalt, concrete, rubber, plastic or geotextile liner material. Manufactured layers are typically found at the ground surface, but also may be buried beneath HTM.

An **agric horizon** forms in response to cultivation and always lies directly below a plowed surface layer (Ap horizon). After precipitation events, turbulent muddy water moves through large pores in the plow layer. Large connecting pores (such as

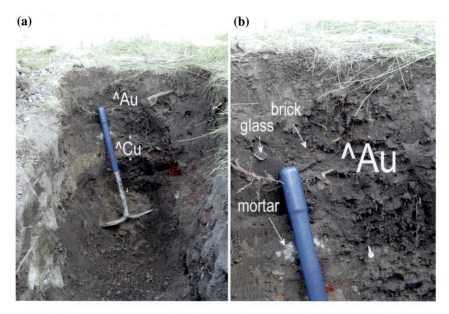

Fig. 6.2 Morphology of anthropogenic soils: **a** Demolition site soil with anthropic epipedon (^Au); **b** Close up of soil shown in (**a**). Photos by J. Howard

wormholes), root channels, and voids between peds directly below the plow layer fill with this turbid water. As it moves into the peds, the suspended material (silt, clay, and humus) is left behind to coat the pore surfaces. The illuvial material is dark-colored, generally brown or dark grayish brown to black. The agric horizon has rarely been identified in the United States. A **calcic horizon** is defined by a significant illuvial accumulation of calcium carbonate. Evidence for the pedogenic accumulation of calcium carbonate ($CaCO_3$) is either the presence of secondary forms, such as masses, threads coatings, pendants, or nodules, or a calcium carbonate equivalent that is higher than that of the underlying horizon. The carbonates typically have moved downward in solution with percolating water, and subsequently precipitated and built up along the wetting front. A calcic horizon is typically designated as Ck. A **cambic horizon** is a subsoil horizon with minimal development. It is typically defined by weak structure, slight reddening, and weak indications of illuvial clay accumulation and/or accumulation of pedogenic carbonate. Under saturated conditions, the cambic horizon has red and gray iron accumulations and depletions (redoximorphic features) as a common characteristic. In semiarid and arid environments, many cambic horizons have redistribution and/or loss of soluble salts. It is typically designated as Bw, Bg, and Bk.

In contrast to free carbonates (i.e. any carbonate that effervesces visibly or audibly when treated with cold dilute HCl), which may be either detrital or pedogenic in origin, **identifiable secondary carbonates** are produced exclusively by chemical weathering, and precipitated from soil solution via pedogenic processes. In urban soils, secondary carbonates may be produced by leaching of carbonate derived from weathered calcareous artifacts. **Lithologic discontinuities** are frequently observed in anthropogenic soils developed in HTM in urban, mine-related and archaeological settings. Lithologic discontinuities are often indicated by differences in color, texture or mineralogy, or by the presence of stone lines or contrasting types of artifacts. Most anthropogenic soils are characterized by horizons with low *n*-values, except where volcanic parent materials are involved. Anthropogenic soils with high *n*-values can be unsuitable for building construction or animal husbandry, and they may indicate a tendency to produce ground subsidence after artificial drainage. Anthropogenic soils on flats are often characterized by **aquic conditions**, especially when they have been compacted excessively by earthmoving equipment. **Anthric saturation** is a special case where aquic conditions have been created intentionally, usually for the agricultural production of rice or cranberries. **Densic materials** occur naturally in materials such as glacial till and volcanic mudflows. However, they can also result from human activities that cause significant compaction of unconsolidated soil material (natural or human-transported). They key characteristic is a severe restriction of plant root penetration. Densic materials are not cemented, so an air-dry fragment will disintegrate in water. **Sulfidic materials**, mainly in the form of pyrite, are most commonly associated with coal-mine and dredge spoil materials, or artificially drained coastal wetland sediments. Upon exposure to aerobic conditions, a **sulfuric horizon** may be formed, because pyrite oxidizes rapidly (within weeks or months) to form sulfuric acid, which in turn may react with carbonate to form extensive amounts of gypsum and perhaps other soluble salts.

Many anthropogenic soils are classified as Entisols, the central concept being a mineral soil with little or no evidence of the development of pedogenic horizons. If horizons are lacking, plant growth is sufficient evidence that the unconsolidated parent material is functioning as soil (Soil Survey Staff 2014). Entisols often have ochric epipedons, and certain types can have albic and/or cambic-like B horizons. Anthropogenic Entisols are typically formed on recently deposited man-made soil materials, or on scalped land surfaces, and are often made by destroying or highly manipulating pre-existing soils with dozers, graders, earthmoving pans, draglines, hydraulic dredging equipment, etc. The earthmoving may take place during surface mining for coal, lignite, sand and gravel, and other natural resources; during the construction of highways, dams and other types of construction in which soil materials are excavated, transported and backfilled; during the burial of wastes at landfills; or during dredging operations. All of these types of operations may form new land surfaces by destroying pre-existing soils, or by depositing artificial fill in low-lying areas. The profile of an anthropogenic Entisol may include buried diagnostic horizons of other soil orders (e.g., calcic, petrocalcic, gypsic, petrogypsic, placic, duripan) if they are buried under recently deposited HAM or HTM to depths of more than 50 cm; or to depths of 30–50 cm, if the thickness of the underlying buried soil is less than twice the thickness of the overlying deposit. Entisols can have any moisture or temperature regime, parent material, vegetation, or age. The only evidence of pedogenic development is generally a small accumulation of organic matter in the upper 25 cm, or a slight illuvial loss of carbonates or clay in the upper 12 cm.

Entisols are soils that basically do not meet the criteria for being classified in other soil orders. They do not have a diagnostic horizon, unless it is a buried horizon, other than an ochric or anthropic epipedon, an albic horizon, or a spodic horizon that has its upper boundary deeper than 2 m. Entisols can have ironstone at any depth, or a salic horizon except that, if the soil is saturated with water within 1 m of the surface for one month or more in some years and has not been irrigated, the upper boundary of the salic horizon must be 75 cm or more below the surface. If the soil is saturated with water within 1 m of the surface for one month or more when not frozen in any part, the sodium adsorption ratio (SAR) may exceed 13% in more than half of the upper 50 cm only if the SAR increases or remains constant with depth below 50 cm. Entisols may also have a histic epipedon if sulfidic material is present within 50 cm of the mineral soil surface. They can have cracks in most years as wide as 1 cm at a depth of 50 cm when not irrigated, but after mixing to a depth of 18 cm, they have less than 30% clay in some subhorizon within a depth of 50 cm, but otherwise do not meet the criteria for classification as a Vertisol. They do not have a sulfuric horizon with an upper boundary within 50 cm of the soil surface, but can have sulfidic material within 50 cm of the mineral soil surface.

Many moderately to well-drained anthropogenic Entisols are members of the Orthent suborder, and associated great groups, depending on moisture regime (Udorthent, Xerorthent, etc.). Anthropogenic Orthents are classified into subgroups using the extragrade formative elements shown in Tables 6.2 and 6.3, which describe their deviation from the typic subgroup. Somewhat poorly to poorly-drained anthropogenic Aquents are subdivided into great groups based on redoximorphic features and type of saturation. **Endosaturation** means the soil is saturated with water in all layers from the upper boundary of saturation to a depth of

Table 6.2 Extragrade formative elements for subgroups of human-altered and human-transported soils (Soil Survey Staff 2014)

Formative element	Description
Anthraquic	Cultivated and flood-irrigated soils with anthric saturation
Anthrodensic	Soils that have a densic contact due to mechanical compaction in more than 90% of the pedon within 100 cm of the mineral soil surface
Anthropic	Soils that have an anthropic epipedon based on the presence of artifacts or midden material
Anthropic Humic	Soils with an anthropic epipedon that also meets the color and total organic carbon requirements of a mollic epipedon throughout
Plaggic	A human-made mineral soil surface layer ≥ 50 cm thick that has been produced by long-term manuring or other additions of organic matter
Haploplaggic	A thinner (25–50 cm thick) plaggen epipedon
Anthroportic	Soils formed in ≥ 50 cm of human-transported material
Anthraltic	Soils formed in ≥ 50 cm of human-altered material

Table 6.3 Examples of subgroups of anthropogenic soils classified using Soil Taxonomy (Soil Survey Staff 2014)

Subgroup	Description
Entisols	
Anthraltic Torriorthent	Soils with little or no evidence of horizonation formed in HAM in an arid setting
Anthraquic Ustorthent	Soils with anthric saturation formed HAM or HTM in an semi-arid setting
Anthrodensic Ustorthent	Soils with a densic contact formed in HAM or HTM in a semi-arid setting
Anthroportic Ustorthent	Soils formed in HTM in an semi-arid setting
Anthraltic Xerorthent	Soils formed in HAM in a Mediterranean climatic setting
Anthraltic Sodic Xerorthent	Soils enriched in sodium formed in HAM in a Mediterranean climatic setting
Anthrodensic Ustorthent	Soils with a densic contact formed in HAM or HTM in a semi-arid setting
Anthrodensic Sodic Ustorthent	Soils enriched in sodium with a densic contact formed in HAM or HTM in an semi-arid setting
Anthroportic Ustorthent	Soils formed in HTM in an semi-arid setting
Anthroportic Endoaquent	Soils with endosaturation formed in HTM
Inceptisols	
Anthraquic Eutrudept	Soils in a humid-temperate climatic setting with incipient horizonation and anthric saturation
Anthroportic Sulfudept	Soils in a humid-temperate climatic setting with a sulfuric horizon formed in HTM

HAM human-altered material; *HTM* human-transported material

200 cm or more from the mineral surface. **Episaturation** is defined as a soil which is saturated with water in one or more layers, but also has one or more unsaturated layers, all within 200 cm of the ground surface. **Anthric saturation** is a human-induced aquic condition found in rice paddy soils.

Cambic and cambic-like horizons have been widely reported to be present in anthropogenic soils. Thus, anthropogenic Entisols grade into soils which meet, or nearly meet, the requirements of the Inceptisol order. Inceptisols have profiles that are more strongly developed than those of the Entisols, but which are too weakly developed to meet the criteria for any of the other soil orders. As a consequence, Inceptisols include a diverse collection of soils. They range from very poorly drained to excessively well drained, and are found in diverse environmental settings. The feature common to all Inceptisols is a relatively weak degree of profile development. Anthropogenic Inceptisols may have ochric, anthropic, plaggen and perhaps other types of epipedons, but they are distinguished from anthropogenic Entisols by having a cambic B horizon. However, there are many exceptions (Soil Survey Staff 2014). Inceptisols can have a wide range in kinds of surface and subsoil horizons. Anthropogenic soils containing calcic or sulfuric horizons are also Inceptisols. Suborders of Inceptisols are defined on the basis of soil moisture regime (Aquepts, Udepts, Ustepts, Xerepts), and classification at the great group level reflects a combination of important properties, including the presence of certain diagnostic horizons (other than the minimally developed ochric epipedon or cambic subsoil horizon), significant levels of sodium and other soluble salts, base saturation level, and pedoturbation. Anthropogenic Inceptisols may be classified into subgroups using the extragrade formative elements shown in Table 6.2, which describe their deviation from the typic subgroup. As discussed further in Chap. 8, some very old anthropogenic soils at archaeological sites are classified as Alfisols and even Ultisols.

Human-altered and human-transported material classes are used to distinguish families of anthropogenic soils within a subgroup. The class names of these differentiae are used to form the family name according to the following sequence: particle size class; human-altered and human-transported material class; mineralogy class; cation-exchange activity class; calcareous and reaction class; soil temperature class (Soil Survey Staff 2014). Anthropogenic soil families are defined using the same categories used for natural soils, except for the human-altered and human-transported material classes (Table 6.4). The control section for these classes is from the soil surface to 200 cm, to the lower boundary of the deepest horizon, or to a lithic or paralithic contact, whichever is shallower. For example, a soil formed in human-transported material that resulted from reclamation of a surface coal mine is: fine-loamy, spolic, mixed, active, calcareous, mesic Anthroportic Udorthent. An example of a soil formed in human-altered material as a result of the mechanical displacement of a pre-existing natric horizon is: fine, araric, smectitic, calcareous, thermic Anthraltic Sodic Xerorthent.

The importance of the concept of the buried soil is especially obvious when classifying anthropogenic soils at the level of the soil series, according to Soil Taxonomy (Soil Survey Staff 2014). If a natural soil is buried beneath a plaggen epipedon, or a capping of HAM or HTM ≥ 50 cm thick, then the pedogenic characteristics of the soil profile developed in the HAM or HTM are used to classify and map the soil. This usually requires the establishment of a new anthropogenic soil series. However, if a natural soil is buried beneath a plaggen epipedon, or a capping of HAM or HTM <50 cm thick, then the characteristics of the buried natural soil are used to classify and map the soil. In this case, the soil may be designated as an anthropogenic phase of an existing soil series.

Table 6.4 HAM and HTM classes used to classify certain anthropogenic soils at the family level in Soil Taxonomy (Soil Survey Staff 2014)

Material Class	Description
Methanogenic	The detectible evolution (>1.6 µg kg^{-1}) of methanethiol odor from the decomposition of non-persistent artifacts (e.g., garbage) or evidence of the collection/burning of methane gas
Asphaltic	A layer or horizon ≥ 7.5 cm thick containing >35 vol.% artifactual asphalt
Concretic	A layer or horizon ≥ 7.5 cm thick containing >35 vol.% artifactual concrete
Gypsifactic	A layer or horizon ≥ 7.5 cm thick containing >35 vol.% artifactual synthetic gypsum products
Combustic	A layer or horizon ≥ 7.5 cm thick containing >35 vol.% artifactual coal cinders
Ashifactic	A layer or horizon ≥ 7.5 cm thick containing >15 number% artifactual coal ash determined by grain count of the 0.02–0.25 mm fraction
Pyrocarbonic	A layer or horizon ≥ 7.5 cm thick containing >5 number% artifacts of pyrolysis (e.g., coke, biochar) determined by grain count of the 0.02–0.25 mm fraction
Artifactic	A horizon ≥ 50 cm thick containing ≥ 35 vol.% cohesive and persistent artifacts
Pauciartifactic	A horizon ≥ 50 cm thick containing ≥ 15 vol.% cohesive and persistent artifacts
Dredgic	A horizon ≥ 50 cm thick with finely stratified (≤ 5 cm thick) HTM deposited by dredging or irrigation
Spolic	A horizon of HTM ≥ 50 cm thick
Araric	A horizon ≥ 7.5 cm thick containing ≥ 3 vol.% mechanically detached and re-oriented pieces of diagnostic horizons or characteristics

For all other soils, no HAM or HTM classes are used

6.3 World Reference Base of Soil Resources

Thirty-two Reference Soil Groups (RSGs) comprise the World Reference Base (WRB) at the highest level of classification (IUSS Working Group 2006, 2015). The second level of classification consists of the RSG name combined with a set of qualifiers and supplementary qualifiers. Ten main diagnostic horizons and materials are used to classify anthropogenic soils according to the WRB (Table 6.5). Many are similar to those used in Soil Taxonomy, and the remainder is generally not known in the United States. Various qualifiers are used to further distinguish RSGs at the second level of classification (Table 6.6).

Anthropogenic soils generally belong to the Anthrosol and Technosol RSGs (Table 6.7). The central concept of an **Anthrosol** is a natural soil that has been drastically modified by long-term human additions of organic and/or inorganic materials. Anthrosols are found at archaeological sites of human habitation (e.g., kitchen middens, graves), and in areas where agricultural activities have been practiced, perhaps during ancient times, for periods of many centuries.

6.3 World Reference Base of Soil Resources

Table 6.5 Selected diagnostic characteristics used to classify anthropogenic soils according to the World Reference Base (IUSS Working Group 2015)

Soil profile characteristic	Description
Diagnostic surface horizons	
Anthraquic	An anthropogenic horizon in paddy soils comprised of a puddled layer and a plow pan
Hortic	Anthropogenic horizon resulting from deep cultivation and long-term additions of organic materials
Irragric	Anthropogenic horizon on raised land built up gradually through continuous additions of irrigation water containing fine particles of inorganic and organic matter
Plaggic	Anthropogenic organic-rich horizon on raised land built up by long-term additions of plaggen
Pretic	Anthropogenic organic-rich horizon resulting from long-term additions of organic materials including biochar
Terric	Anthropogenic horizon on raised land built up by long-term additions of inorganic materials
Diagnostic subsurface horizons	
Hydragric	Anthropogenic horizon in paddy soils characterized by redoximorphic features
Thionic	Extremely acid horizon produced by weathering of sulfides due to strip mining or artificial draining of swamp soils
Other diagnostic materials	
Artifacts	Solid or liquid substances that are created or modified by human manufacturing or artisanal activities
Technic Hard Rock	Human-manufactured, consolidated material used to seal the ground surface

A **Technosol** is characterized by the presence of materials manufactured, altered or exposed by human technology that otherwise would not be present at the Earth's surface. Included in the Technosol RSG are urban soils associated with residential and industrial landscapes that contain abundant technogenic artifacts, manufactured (paved) land, and mine soils. In the case where a natural soil is buried by human-deposited material, the overlying material and the buried soil are classified as one soil if both together qualify as an Anthrosol or a Technosol. Otherwise, the overlying material is classified with preference if it is ≥ 50 cm, and the buried soil is classified separately and recognized with the Thapto (buried) prefix (Rossiter 2007). For example, Spolic Technosol (Thapto-Luvisolic) denotes a soil formed in rubble-bearing fill overlying a buried soil formed in loess. In all other cases, the buried soil is classified with preference.

Because Anthrosols were formed primarily as a result of human additions to natural soils over long time periods, or by modifying natural soils through irrigation or artificial ponding, they were formed in HAM as defined in the U. S. Soil Taxonomy. Anthrosols are essentially soils that have a hortic, irragric, plaggic, pretic, or terric horizon ≥ 50 cm thick, or an anthraquic horizon and an underlying

Table 6.6 Selected qualifiers used to classify anthropogenic Reference Soil Groups in the World Reference Base (IUSS Working Group 2015)

Formative Element	Description
Anthraquic	Having an anthraquic horizon
Densic	Having artificial compaction within 50 cm of the ground surface to the extent that roots cannot penetrate
Ekranic	Having technic hard material starting at ≤ 5 cm from the soil surface
Garbic	A Technosol having a layer ≥ 20 cm thick within 100 cm of the surface containing $\geq 20\%$ by volume artifacts containing $\geq 35\%$ by volume organic waste materials
Hortic	Having a hortic horizon
Hydragric	Having a hydragric horizon
Irragric	Having an irragric horizon
Plaggic	Having a plaggic horizon
Reductic	Having reducing conditions in $\geq 25\%$ of the soil volume caused by gaseous emissions of CH_4 or CO_2
Spolic	A Technosol having a layer ≥ 25 cm thick within 100 cm of the surface containing >20% by volume artifacts of industrial waste (mine spoil, dredgings, rubble, etc.)
Technic	Having $\geq 10\%$ by volume artifacts in the upper 100 cm
Transportic	Having a layer ≥ 30 cm thick comprised of solid or liquid material moved from a source outside the immediate vicinity by human activity
Urbic	Having a layer ≥ 30 cm thick within 100 cm of the surface containing $\geq 20\%$ by volume artifacts of rubble and refuse from human occupation

hydragric horizon with a combined thickness of ≥ 50 cm. Plaggic and Terric Anthrosols are extensive in northwestern Europe, especially in the Netherlands. Irragric Anthrosols are found in dry climate soils impacted by irrigation, as in Mesopotamia and the Indus Valley, as well as at desert oases. Hydragric Anthrosols are mainly rice paddy soils which occupy extensive parts of China and southeastern Asia. The hortic horizon is similar to the anthropic epipedon, as originally defined in Soil Taxonomy (Soil Survey Staff 1975), hence Hortic Anthrosols are characteristic of kitchen middens and garden soils at archaeological sites. Pretic Anthrosols include Terra Preto de Indio (Amazonian Dark Earth) soils in the Amazon River basin of Brazil and surrounding areas, and the highly productive soils of the Aztec civilization. In contrast to Anthrosols, which are metagenetic, Technosols are neogenetic anthropogenic soils whose pedogenic characteristics reflect the influence of modern human technology. Technosols are basically anthropogenic soils formed in HTM that contain $\geq 20\%$ artifacts by volume in the upper 100 cm, or which have technic hard material (manufactured layer) at, or near, the ground surface.

6.3 World Reference Base of Soil Resources

Table 6.7 Classification and characteristics of anthropogenic soils according to the World Reference Base (Modified from IUSS Working Group 2015)

WRB classification		Anthropogenic Soil Type	Origin	Description	Type of parent material
Anthrosol	Hortic	Garden soils; Hortisol; Fimic Anthrosol	Long-term cultivation and/or human additions of manure or organic kitchen wastes	Thick, black, organic-rich hortic *A horizon, containing high levels of phosphorous and biological activity, overlying agric horizon	HAM
	Hortic	Midden-soils	Long-term human occupation at archaeological sites; disposal of food wastes	Artifact-bearing, thick, black, organic-rich, hortic *A horizon, possibly containing high levels of phosphorous and other nutrients	HAM
	–	European Dark Earth (Paleosol)	Long-term human occupation and horticultural activities at Medieval European archaeological sites	Artifact-bearing, thick, black, organic-rich, paleosols (*A horizons?) possibly containing high levels of phosphorous and other nutrients	HAM and HTM?
	Plaggic	Plaggen soils, Esch soils	Plaggen agriculture at Medieval European archaeological sites; long-term human additions of plaggen (sod mixed with animal manure)	Thick, sandy, brown or black, artifact-bearing plaggic ^A horizon found on raised land overlying remnants of original subsoil	HTM
	Terric	Earth-cumulic and mud-cumulic soils	Long-term human additions of inorganic materials (loess, mud, marl, etc.)	Thick, terric ^A horizon on raised land with color inherited from inorganic materials added; possibly highly calcareous	HTM
	Pretic	Terra Preta de Indio; Amazonian Dark Earth	Long-term human additions of organic kitchen wastes and biochar to soils in the Amazon River basin of South America	Thick, brown or black, organic-rich, artifact-bearing pretic *A horizon containing abundant charcoal (biochar) fragments and showing high degree of biological activity	HAM

(continued)

Table 6.7 (continued)

WRB classification		Anthropogenic Soil Type	Origin	Description	Type of parent material
	Irragric	Irrigation soils: Oasis soils	Long-term additions of irrigation water	Thick, irragric *A horizon on raised land enriched in clay, organic matter, and biological activity	HAM
	Hydragric	Paddy soils	Long-term rice cultivation and human additions of organic wastes in flooded paddy field	Puddled layer and underlying plowpan (anthraquic *A horizon) overlying zone with common redoximorphic features (hydragric *B horizon)	HAM
	Necric[a]	Grave soils	Human interment in grave shaft	Grave soils with a subsurface residual accumulation of organic matter derived from decomposed corpse, associated with artifacts indicative of human interment	HAM
Technosol	Urbic	Residential urban soils	Modern sites of human habitation in cities	Technogenic artifact assemblage comprised of building rubble and artificial objects indicative of human habitation	HTM
	Spolic	Coal mine soils	Modern strip-mining operations	Technogenic artifact assemblage comprised of mine spoil; includes hyperskeletic soils	HTM
	Spolic	Burial mound soils	Ancient ceremonial burial sites	Technogenic artifact assemblage comprised of human-deposited soil or sediment and grave goods	HTM
	Spolic	Industrial urban soils	Modern sites of industrialization	Technogenic artifact assemblage comprised of industrial wastes including building rubble, mine spoil, dredgings, slag, cinders, or ash; includes hyperskeletic soils	HTM

(continued)

6.3 World Reference Base of Soil Resources

Table 6.7 (continued)

WRB classification		Anthropogenic Soil Type	Origin	Description	Type of parent material
	Garbic	Landfill soils	Modern sites of human habitation in cities	Technogenic artifact assemblage comprised of organic wastes; includes soils with reductic and linic characteristics	HTM
	Ekranic	Sealed soils; technic hard material; manufactured land	Modern sites of human habitation in cities	Soils sealed by, or containing, one or more layers of technic hard material; includes certain soils with isolatic and leptic characteristics	–
	Isolatic	Rooftop soils; Edifisols	Modern sites of human habitation in cities	Soils in pots or rooftop gardens	HTM

[a]Term used informally

U.S. Soil Taxonomy terms: *HAM* human-altered material (*); *HTM* human-transported material (ˆ)

References

Alexandrovskaya EI, Alexandrovskiy AL (2000) History of the cultural layer in Moscow and accumulation of anthropogenic substances in it. Catena 41:249–259

Bryant RB, Galbraith JM (2003) Incorporating anthropogenic processes into soil classification. In: Eswaran H, Rice T, Ahrens R, Stewart BA (eds) Soil classification—a global desk reference. CRC Press, Boca Raton, FL, pp 57–66

Buol SW, Southard RJ, Graham RC, McDaniel PA (2011) Soil genesis and classification. Wiley-Blackwell, West Sussex, UK 543 pp

Burghardt W (1994) Soils in urban and industrial environments. Z Pflanz Bodenk 157:205–214

Capra GF, Ganga A, Grilli E, Vacca S, Buondonno A (2015) A review on anthropogenic soils from a worldwide perspective. J Soils Sediments 15:1602–1618

Charnzynski P, Bednarek R, Greinert A, Hulisz P, Uzarowicz L (2013) Classification of technogenic soils according to World Reference Base in the light of Polish experiences. Soil Sci Annu 64:145–150

Dazzi C, Papa GL, Palermo V (2009) Proposal for a new diagnostic horizon for WRB anthrosols. Geoderma 151:16–21

Eswaran H, Rice T, Ahrens R, Stewart BA (2003) Soil classification—a global desk reference. CRC Press, Boca Raton, FL 263 pp

Fanning DS, Fanning MCB (1989) Soil morphology, genesis, and classification. Wiley, New York 395 pp

Gong A, Zhang G (2007) Chinese soil taxonomy: a milestone of soil classification in China. Sci Found China 15:41–45

Gong ZT, Zhang GL, Chen ZC (2003) Development of soil classification in China. In: Eswaran H, Rice T, Ahrens R, Stewart BA (eds) Soil classification—a global desk reference. CRC Press, Boca Raton, FL, pp 101–125

Gray JM, Humphreys GS, Deckers JA (2011) Distribution patterns of World Reference Base soil groups relative to soil forming factors. Geoderma 160:373–383

Isbell R (2016) The Australian soil classification, 2nd edn. CSIRO Publishing, 152 pp

IUSS Working Group (2006) World Reference Base for soil resources 2006: World Soil Resources Report 103. Food and Agriculture Organization United Nations, Rome, Italy 145 p

IUSS Working Group WRB (2015) World Reference Base for soil resources 2014, update 2015 International soil classification system for naming soils and creating legends for soil maps, World Soil Resources Reports No. 106. FAO, Rome

Krasilnikov P, Arnold RW, Ibanez JJ (2010) Soil classifications: their origin, the state-of-the-art and perspectives. World Congress of Soil Science, Solutions for a changing world. Brisbane Australia, 19–22

Lebedeva II, Gerasimova MI (2012) Diagnostic horizons in the Russian soil classification system. Eurasian Soil Sci 45:823–833

Lehmann A, Stahr K (2007) Nature and significance of anthropogenic urban soils. J Soils Sediments 7:247–260

Meuser H, Blume HP (2001) Characteristics and classification of anthropogenic soils in the Osnabruck area, Germany. J Plant Nut Soil Sci 164:351–358

Prokofyeva TA, Martynenko IA, Ivannikov FA (2011) Classification of Moscow soils and parent materials and its possible inclusion in the classification system of Russian soils. Eurasian Soil Sci 44:561–571

Puskas I, Farsang A (2009) Diagnostic indicators for characterizing urban soils of Szeged, Hungary. Geoderma 148:267–281

Rossiter DG (2007) Classification of urban and industrial soils in the World Reference Base for soil resources. J Soils Sediments 7:96–100

Sarmast M, Farpoor MH, Boroujeni IE (2016) Comparing Soil Taxonomy (2014) and updated WRB (2015) for describing calcareous and gypsiferous soils, central Iran. Catena 145:83–91

References

Shi XZ, Yu DS, Warner ED, Sun WX, Petersen GW, Gong ZT, Lin H (2006) Cross-reference system for translating between genetic soil classification of China and Soil Taxonomy. Soil Sci Soc Am J 70:78–83

Simonson RW (1962) Soil classification in the United States. Science 137:1027–1034

Soil Survey Staff (1975) Soil taxonomy: a basic system of soil classification for making and interpreting soil surveys. USDA-SCS Agric. Handbook 436, U.S. Government Printing Office, Washington, DC

Soil Survey Staff (2011) Buried soils and their effect on taxonomic classification. Soil Survey Technical Note no. 10, USDA, NRCS, 7 pp

Soil Survey Staff (2014) Keys to soil taxonomy, 12th edn. U.S. Department of Agriculture, Natural Resources Conservation Service, 372 pp

Chapter 7
Anthropogenic Soils in Agricultural Settings

Anthrosols are the characteristic type of anthropogenic soil found wherever people have practiced agriculture for many centuries. They were generally formed as a result of ancient methods used to improve soil fertility and productivity. Six basic types of Anthrosol are recognized in the WRB: (1) Hortic, (2) Plaggic, (3) Terric, (4) Pretic, (5) Irragric, and (6) Hydragric. Hortic, Plaggic and Pretic Anthrosols are natural soils that were modified by long-term human additions of organic materials, whereas Terric Anthrosols were formed by additions of mainly inorganic earth materials (e.g., marl, loess, mud). Irragric Anthrosols have an anthropogenic mineral surface horizon formed from a pre-existing natural soil by the gradual buildup of fine silt and clay, and possibly fertilizers, soluble salts, organic matter, as a result of irrigation. Hydraquic Anthrosols are rice paddy soils which generally consist of a cultivated plow layer and an underlying plowpan, which make up an anthraquic horizon, overlying a gleyed, illuvial subsoil comprising a hydragric horizon. The upper part of the anthraquic horizon is enriched in organic matter as a result of human additions of manure, and the lower part is densic with platy structure. The hydragric horizon is characterized by redoximorphic features because of repeated wetting and drying. Hortic, Plaggic and Pretic Anthrosols are characterized by a thick, black, artifact-bearing, organic-rich topsoil characterized by elevated levels of P and other plant nutrients. Hortic and Pretic Anthrosols show a high degree of biological activity, both in terms of macroafauna and microfauna. Plaggic, Terric and Irragric Anthrosols are found on land surfaces which have been raised by human additions of organic or inorganic earth materials. Hortic, Plaggic and Pretic Anthrosols often contain abundant charcoal fragments, but Pretic Anthrosols (Terra Preta de Indio) are distinguished by the presence of biochar and a tropical setting in the Amazon River basin. Hydragric Anthrosols are typically found in excessively wet climatic settings, whereas Irragric Anthrosols are found in dry climate regions, and have unique pedogenic features associated with excess soluble salts. Most of these soils are not addressed in U.S. Soil Taxonomy because they are not known to exist in the United States, or are of very limited extent.

7.1 Introduction

Approximately 13% of the Earth's surface has been affected by agriculture, hence cultivation of the land for food production has created the most extensive terrains of anthropogenic soils. The earliest archaeological evidence for agricultural activities dates to the early Holocene about 9000 BC, and includes artifacts found in Iraq that were used for tilling and harvesting. There is also archaeological evidence in southern Iraq that ancient Sumerians were digging irrigation canals and ditches by about 7500 BC (Brevik and Hartemink, 2010). Rice and other crops were being cultivated in India and China by ~7000 BC, and sedentary agricultural activities were occurring in parts of Europe by 5000 BC. Agricultural activities were well underway in ancient Greece, and the Chinese had developed the first system of soil classification used for taxation, by about 2000 BC. Farming was being carried out by the Roman Empire in the Mediterranean region and Europe, by the Hohokam Indians in North America, the Maya and Aztecs in Central America, and the Amazonians in South America around 500 BC. The Aztec civilization had ended by about AD 700, but farming continued in North and South America until the European conquest beginning about AD 1500. Following the fall of Rome in AD 410, much of the early soils knowledge gained by the Greeks and Romans was lost or forgotten.

The first quasi-scientific studies of soils did not begin in Europe until the 16th and 17th centuries. These early studies were focused on agriculture, and led to the recognition of the importance of water, nitrogen, carbon and phosphorous, cycling of inorganic elements as nutrients, soil structure, and the role of humus, in soil fertility and plant growth. However, these studies were largely based on observations of nature. Truly scientific experimentation did not begin until the late 18th century, and by the mid-19th century, most of the essential plant nutrients had been recognized, along with the utility of mineral fertilizers. The modern fields of soil chemistry and soil fertility reached a high degree of development during the 20th century.

It is clear from the historical and archaeological record that ancient civilizations had acquired a great deal of early knowledge about soils and agriculture long before the modern advances of the 19th and 20th centuries. Writings on clay tablets in Mesopotamia dating back to 2500 BC mentioned the fertility of land, and recorded the yields of barley and other crops over time. They also recorded the declining yields of the Sumerians, and later the Babylonians, which resulted from salinization brought on by irrigation. Early knowledge about soils and agriculture no doubt was acquired purely by trial and error. It has been recognized for thousands of years that certain soils are inherently infertile, and will fail to produce satisfactory yields when continuously cropped. Simple field observations, such as the effect dead bodies had on increasing the growth of plants, probably led to the recognition that soil fertility could be improved by human additions of various types of amendments. This observation dates back at least to Archilochus in ancient Greece about 700 BC, and similar references in the Old Testament may date from before 1000 BC.

7.1 Introduction

Perhaps the oldest known agricultural technique is manuring, i.e. the human addition of organic matter in the form of animal feces. Manuring generally involves the collection, storage, and application of animal wastes onto the ground surface, after which the manure is plowed into the surface soil. In the case of "green manure" plant stubble, often in the form of a cover crop, is plowed into the topsoil. When manuring was first used is unknown, but it is commonly used in paddy field cultivation of rice today following traditional methods which have been followed for thousands of years. It potentially dates to the early Holocene, based on genetic evidence showing that domesticated rice originated from wild rice before 6200 BC. The earliest paddy field archaeological site dates to around 4330 BC, based on radiocarbon dating of rice grains, although other sites have been tentatively dated to around 7000 BC. The use of manure was part of plaggen agriculture, which dates to about 1000 BC in northern Europe, and which was being widely used in the southern Europe by 500 BC. There are references to manure heaps in the ancient Greek poem "The Odyssey" dating to around 900 BC, and in writings from around 300 BC, the Greek philosopher Theophrasus recommended the abundant manuring of certain soils. He endorsed the practice of stabling cattle, and the use of plant materials (e.g., sod or straw) as bedding to collect wastes in an animal stall, which was thought to increase the fertility of the manure. Theophrasus also developed a ranking of animal manures according to their richness, which he listed in order of decreasing value: human > swine > goat > sheep > cow > oxen > horse (Tisdale et al. 1985). The ancient Romans had a similar list, but ranked bird guano as superior to human excrement. Theophrasus mentions the value of saltpeter (KNO_3) as useful for fertilizing plants, and noted that some farmers obtained benefits from plowing the stubble from certain bean crops (legumes) into the topsoil. He also suggested that adding fertile soil materials to an infertile soil could lead to increased productivity, and beneficial changes in certain physical characteristics.

The ancient Chinese are known to have added human excrement, mud and various types of earth materials to croplands before about 1000 BC (Gong et al. 1999). The ancient Minoans dug up marl and applied it to crops perhaps as early as ∼2000 BC, and a similar practice was followed by the ancient Greeks and Romans, who classified various liming materials according to the utility for various crops. The Romans also recognized the utility of adding wood ashes as an agricultural amendment, which they plowed into the soil. Ancient cultures recognized the value of burning crop stubble to generate wood ash and control weeds. Many early writers believed that soil color was a useful criterion for identifying fertile soils, brown or gray soils being generally less fertile than black ones. However, around AD 50, the Roman historian Columella noted the infertility of black marshland soils. He recommended a "taste test" to assess the degree of acidity and salinity of soils, and suggested that the best test for the suitability of a soil for agriculture was simply whether or not a certain crop would grow sufficiently. The use of organic and other amendments in ancient agriculture appears to have been practiced in many parts of the world especially from about 500 BC to the end of the Middle Ages, as shown by the wide distribution of plaggen and paddy soils, Terra Preta de Indio, European Dark Earth soils, and other anthropogenic soils.

7.2 Types of Anthrosols

Anthropogenic soils in agricultural settings comprise the Anthrosol Reference Soil Group in the World Reference Base (IUSS Working Group 2015). Anthrosols are found wherever people have practiced agriculture for generations, often for many centuries or even millennia. Most of these soils are not addressed in U.S. Soil Taxonomy because they are not known to exist in the United States, or are of very limited extent. Hence, the terminology of the WRB is followed here.

There are six basic types of Anthrosol: (1) Hortic, (2) Plaggic, (3) Terric, (4) Pretic, (5) Irragric, and (6) Hydragric (Table 7.1). All are natural soils that have

Table 7.1 Classification and characteristics of Anthrosols found in agricultural settings

Anthrosol type	Also known as	Origin	Description
Hortic	Garden soils; Hortisols, Fimic Anthrosols; Midden soils	Long-term cultivation and/or human additions of manure or, organic kitchen wastes	Thick, black, organic-rich hortic *A horizon, containing high levels of phosphorous and biological activity, overlying agric horizon
Plaggic	Plaggen soils, Esch soils, Enk earth soils	Long-term human additions of plaggen (sod mixed with animal manure)	Thick, sandy, brown or black plaggic ^A horizon found on raised land overlying remnants of original subsoil
Terric	Earth-cumulic and mud-cumulic soils	Long-term human additions of inorganic materials (loess, mud, marl, etc.)	Thick terric ^A horizon on raised land with color inherited from inorganic materials added; possibly highly calcareous
Pretic	Terra Preta de Indio; Amazonian Dark Earth	Long-term human additions of organic kitchen wastes, biochar and charcoal	Thick brown or black, organic-rich pretic *A horizon containing abundant charcoal fragments and showing high degree of biological activity
Irragric	Irrigation soils; Oasis soils	Long-term additions of irrigation water	Thick irragric *A horizon on raised land enriched in clay, organic matter, and biological activity
Hydragric	Paddy soils	Long-term rice cultivation and human additions of organic wastes in flooded paddy field	Puddled layer and underlying plowpan (anthraquic horizon) overlying zone with common redoximorphic features (hydragric horizon)

been strongly altered by human activity, usually by forming a new, overthickened A horizon, which rests on subsoil remnants of the original profile. Hortic, Plaggic and Pretic Anthrosols are comprised of a thick, black (10YR2/1), organic-rich topsoil characterized by elevated levels of phosphorous, and which commonly contains artifacts of archaeological significance. Hortic and Pretic Anthrosols show a high degree of biological activity, both in terms of macrofauna and microfauna. Plaggic, Terric and Irragric Anthrosols are all found on raised land surfaces, but Plaggic and Terric Anthrosols are on surfaces which have been raised directly by human additions of solid earth materials, whereas Irragric Anthrosols are on ground surfaces which have been raised indirectly as a result of irrigation. Hortic, Plaggic and Pretic Anthrosols often contain abundant charcoal fragments, but Pretic Anthrosols are distinguished by the presence of biochar. Hydragric Anthrosols are characterized by redoximorphic features produced by prolonged submergence beneath water ponded for rice farming, typically in an excessively wet climatic setting. In contrast, Irragric Anthrosols are found in dry climate regions, and have unique pedogenic features associated with excess soluble salts. Anthrosols are formed in human-altered material, as defined in U.S. Soil Taxonomy, except for Plaggic and Terric Anthrosols which are formed in human-transported material.

The key characteristics which distinguish the basic types of Anthrosols according to the WRB are summarized in Table 7.2. However, note that certain Anthrosol types are defined as much by geocultural setting as by pedological features. For example, Plaggic Anthrosols are generally restricted to northern Europe, and are best known in Belgium, the Netherlands and northwestern Germany. Pretic Anthrosols are essentially restricted to the Amazon River basin of Brazil and Colombia. There is also considerable overlap between certain types of Anthrosols associated with agricultural settings, and those of archaeological interest, which are discussed further in Chap. 8.

7.3 Hortic Anthrosols

The central concept of the Hortic Anthrosol in the WRB is based on the Chinese paradigm of a Fimic Anthrosol (Gong et al. 1999). It is a natural soil of any type which has been modified drastically by long-term human additions of organic materials used as fertilizer for vegetable gardening, and other horticultural activities. Applications of organic amendments and cultivation over a protracted period of time, typically on a scale of many centuries, results in the eventual development of an anthropogenic soil profile comprised of a hortic surface horizon (fimic epipedon), and an underlying agric horizon (Fig. 7.1). The hortic horizon is a thick, black, organic-rich layer characterized by high levels of extractable phosphorous. These characteristics also match those of the anthropic epipedon, as originally conceived in the U.S. system of soil taxonomy (Soil Survey Staff 1975; Smith 1986). Hence, Hortic Anthrosols include soils formed at archaeological sites of human occupation (middens) by the disposal of organic kitchen wastes (discussed

Table 7.2 Key properties distinguishing different types of Anthrosols, according to the World Reference Base (IUSS Working Group 2015)

WRB diagnostic criteria	Type of diagnostic horizon					
	Hortic	Plaggic	Terric	Pretic	Irragric	Hydragric
Color (moist)	Chroma ≤ 3	Chroma ≤ 4	Inherited	Chroma ≤ 3	–	–
Organic carbon	$\geq 1\%$	$\geq 0.6\%$	–	$\geq 1\%$	$\geq 0.5\%$	–
Exch. Ca + Mg	–	–	–	≥ 2 cmol kg^{-1}	–	–
Base saturation	$\geq 50\%$	<50%	$\geq 50\%$	–	–	–
Phosphorous	≥ 100 mg kg^{-1}	Possibly high	–	≥ 30 mg kg^{-1}	–	–
Thickness	≥ 20 cm	≥ 20 cm	≥ 20 cm	≥ 20 cm	≥ 20 cm	≥ 10 cm
Biol. activity	$\geq 25\%$ (vol)	–	Possibly high	<25% (vol)	$\geq 25\%$ (vol)	–
Artifacts	Common	<20%	Common	$\geq 1\%$	Common	–
Carbonates	Possibly high	–	Possibly high	–	Possibly high	–
Raised land	No	Yes	Yes	No	Yes	No
Charcoal	–	–	–	$\geq 1\%$	–	–
Redox features	–	–	–	–	–	Common

Note phosphorous measured by different methods

7.3 Hortic Anthrosols

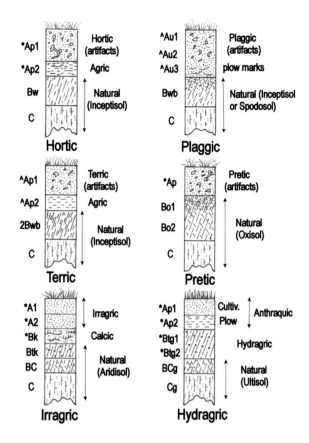

Fig. 7.1 Soil profile characteristics of different types of Anthrosols

further in Chap. 8). Hortic Anthrosols are found worldwide especially in, and around, cities whose histories date back for hundreds or thousands of years (Fig. 7. 2). In China, there are an estimated 0.7–1.0 million ha of Hortic Anthrosols (Gong et al. 1999). Agric horizons are perhaps best known in China, Belgium, and India.

A key characteristic of a Hortic Anthrosol is the hortic horizon (Latin, *hortus*, garden), defined as a mineral surface horizon created by deep cultivation, intensive fertilization, and/or long-continued applications of human and animal wastes, and other organic residues. A hortic horizon has the following properties (IUSS Working Group 2015): (1) A Munsell color value and chroma ≤ 3, (2) a weighted average of $\geq 1\%$ organic carbon, (3) 0.5 M $NaHCO_3$-extractable P_2O_5 of ≥ 100 mg kg^{-1} fine earth in the upper 25 cm, (4) a BS $\geq 50\%$ (1 M Ammonium acetate), (5) $\geq 25\%$ by volume evidence of animal activity, and (6) a thickness of ≥ 20 cm. In order to be classified as a Hortic Anthrosol, the hortic horizon must be ≥ 50 cm thick.

Well drained, sandy Hortic Anthrosols also may be characterized by an agric horizon, i.e. an anthropogenic illuvial horizon lying directly beneath a plowed A

Fig. 7.2 Hortic Anthrosol dating to the 13th–18th centuries in Leicester, England. Soil profile is very homogeneous, and 1–2 m thick, lying beneath ~75 cm of modern overburden. Photo by University of Leicester Archaeological Services

horizon. An agric horizon is ≥ 10 cm thick with illuvial humus, silt and clay comprising ≥ 5% of the horizon by volume. It is typically characterized by abundant earthworm burrows (krotovina) and clay lamellae. This horizon develops because disturbance by plowing triggers the translocation of fine particles into the zone immediately underlying the plow layer. Over time, illuvial particles form lamellae, or coatings in worm burrows or on ped surfaces.

Some vegetable gardens in China date back more than 1000 years, where Hortic Anthrosols have formed under an udic soil moisture regime as a result of frequent watering, intensive cultivation, and applications of night soil (human excrement), animal manure, yard wastes, coal cinders, ashes, and miscellaneous organic kitchen refuse (Gong et al. 1999). Garden soils with hortic horizons about 22 cm thick, with an organic carbon content of ~3%, have formed in 20–30 years (Zhang et al. 2003), and those gardened for centuries are often 50–80 cm or more in thickness. Available P content is exceptionally high, even exceeding that of guano, as a result of vegetable cultivation; lower levels are observed in grain crop soils. Original soil characteristics may influence the properties (e.g., pH) of the resulting anthrosol, but hortic horizons generally have elevated nutrient status, pH, and carbonate content, compared with the natural soils from which they were formed. They have excellent agricultural properties such as strongly developed structure, high porosity, high water retention capacity, and enhanced soil biological characteristics. Hortic horizons are characterized by extensive earthworm activity with earthworm casts and burrows typically comprising >25% by volume, and with well developed microbial populations dominated by bacteria, which increase with increasing stage of ripening. Samples of 86 hortic horizons from Chinese anthrosols had average levels of $NaHCO_3$-extractable, citric acid-extractable, and total, phosphorous of 282, 1844 and 3610 mg kg^{-1}, respectively (Gong et al. 1997). Levels increased dramatically with increased ripening over time (Fig. 7.3).

The extent of biological activity that may occur in a Hortic Anthrosol is illustrated by a 27 cm-thick hortic horizon developed in a 100 year-old garden soil in

7.3 Hortic Anthrosols

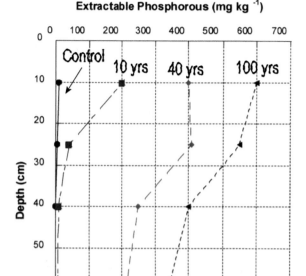

Fig. 7.3 Extractable phosphorous (0.5 M NaHCO$_3$) depth profiles in Hortic Anthrosols of different ages in China (After Gong et al. 1997). Phosphorous levels progressively increase in soils after 10, 40 and 100 years of continuous vegetable cultivation compared to soil with no vegetable cultivation (control)

Detroit, Michigan, USA (Fig. 7.4), which formed partly under an artificial irrigation system that operated from ~1918 to 1988. The hortic horizon was black (Table 7.3), and had an organic carbon content of ~3.5% (Howard et al. 2013). Abundant artifacts reflecting human habitation were concentrated at the base of the hortic horizon, including a copper penny dated 1920. The soil is characterized by abundant vertical burrows formed by the earthworm *Lumbricus terrestris*, an invasive species which was introduced to southeastern Michigan by European settlers during the 18th century. A count showed a population density of 150 worms m^{-2}, which is comparable to that of forest soils. The hortic horizon is interpreted as a biomantle (Fig. 7.5), hence the concentration of artifacts at depth is attributed to burial of objects formerly on the ground surface by earthworm casting and upbuilding (Howard et al. 2015). Roman coins in English soils were observed to have been buried by the earthworm casting activity to a similar depth within ~30 years (Darwin 1881).

Three Hortic Anthrosols developed in convent garden soils in Poland (Table 7.4), and cultivated for ~300–700 years under an udic soil moisture regime, had hortic horizons 93–160 cm thick, compared with the 24 cm-thick Ap of the local native soil (Gasiorek and Niemyska-Lukaszuk 2008). The hortic horizons had organic carbon levels of 2.5–3.9%, compared with 1.5% in natural soils. They also had total N levels that were twice as high as those in natural soils. Although C/N ratios were similar to those of natural soils, humin and humic acid

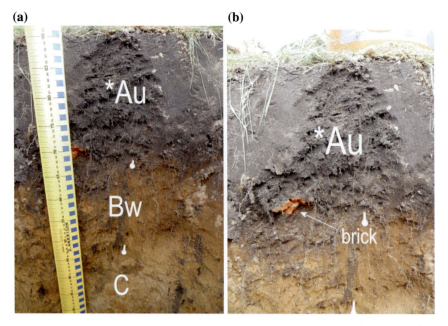

Fig. 7.4 One hundred year-old garden soil with hortic (*Au) horizon at an historic archaeological site in Detroit, Michigan. Note extensive biological activity indicated by *vertical burrows* of the earthworm *Lumbricus terrestris*, and invasive species brought to North America by French and British settlers during the 18th century. Scale in cm. Photo by J. Howard

predominated over fulvic acid in the anthropogenic soils. Three soils in northern Russia which had been fertilized with a mixture of peat and animal manure, and farmed for an estimated 300–800 years, had *A horizons that were 59 cm or more in thickness (Hubbe et al. 2007). The characteristics of a representative profile are shown in Tables 7.5 and 7.6. These soils had pH values ranging from 5.9 to 7.1, and a base saturation of 100%. Citric acid-extractable P ranged from 340 to 790, and total P levels were 710–1410, mg kg^{-1}. Organic carbon is 3.7% in the representative profile, and decreases to 1.5% at the base of the *A horizon. Hence, these soils meet the requirements of a Hortic Anthrosol, according to the current WRB (Table 7.2).

Sandor (1993) studied the effects of cultivation on soils at two ancient agricultural sites (Table 7.7). He compared cultivated and uncultivated soils at a semi-arid archaeological site in southwestern New Mexico ($\sim$50 cm/year annual precip.). The soils were estimated to have been cultivated for $\sim$500 years, and were abandoned for about 800 years. The cultivated soils had thicker A horizons, ranging up to 60 cm thick, with an organic carbon content of $\sim$1%. The cultivated topsoils were lighter in color, massive, and showed a loss of N, P and C, whereas bulk density increased, compared with uncultivated soils. This type of degradation is also commonly seen in modern agricultural soils. In contrast, ancient Peruvian agricultural soils in an arid setting ($\sim$10 cm/year annual precip.) were thicker and

7.3 Hortic Anthrosols

Table 7.3 100 year-old Anthropic Udorthent with hortic horizon at an historic archaeological site in Detroit, Michigan

Horizon	Depth (cm)	Description
*Au	0–27	Black (10YR2/1) sand to loamy sand; medium fine granular (0–16 cm) to massive (16–27 cm); very friable, non-sticky, non-plastic; many fine roots, very few coarse roots; common artifacts (5–10%) at 23–27 cm including glass, wood, nails, ceramics, two copper pennies dated 1878 and 1920; brick, mortar; weak reaction; abrupt smooth boundary. Hortic horizon
Bw	27–49	Strong brown (7.5YR4/6) fine sand to loamy sand; massive to single grain; very friable, non-sticky, non-plastic; few fine roots; sparse artifacts (<5%); 0.5–1.5 cm in diameter; common *Lumbricus terrestris*; many medium krotovina; weak reaction; gradual wavy boundary. Weakly weathered brick and coal cinders, significant weathering of cement and iron objects; strong reaction; clear smooth boundary
C	49–150	Yellowish-brown (10YR5/6) with common coarse, distinct strong brown (7.5YR5/6) and few, fine, prominent yellowish-red (5YR5/8) mottles (Fe-oxide proto-nodules) sand; massive to single grain; very friable, non-sticky, non-plastic; very few fine roots; few coarse to very coarse krotovina to ~100 cm; very weak reaction; abrupt smooth boundary

showed higher levels of C, N and P, and no difference in bulk density, compared with local uncultivated soils. These soils had been cultivated for about 1500 years, and had been abandoned for about 500 years. These results are interesting because they suggest that agricultural soils initially undergo a period of degradation, for as long as 500 years, and that these effects persist when abandoned for a long period of time. In contrast, cultivation of soils for very long time periods, perhaps on the order of 1000 years or more, can leave long-lasting positive effects on soils, even when abandoned for 500 years. Furthermore, relatively little degradation seems to have occurred after abandonment. Davis et al. (2000) also studied some ancient agricultural soils in an arid setting in Arizona, USA (Table 7.8). These soils showed signs of increased salinity as a result of cropping, indicated by electrical conductivities (EC) >4 dS m^{-1}, and sodium adsorption ratios (SAR) >13.0. This is a common problem in modern agriculture, and is discussed further in Sect. 7.6.

Soils classified as Hortic Anthrosols in the WRB would probably be classified mainly as Anthropic Entisols in U.S. Soil Taxonomy. As discussed below, midden and European Dark Earth soils at archaeological sites are probably classified mainly as Hortic Anthrosols. Also included are probably some ancient agricultural soils amended with a mixture of peat and manure, which were previously classified as Plaggic or Terric Anthrosols (e.g., Conry 1974; Hubbe et al. 2007).

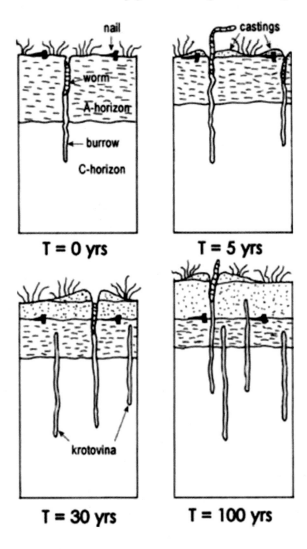

Fig. 7.5 Interpretation of how the casting activity of the earthworm *Lumbricus terrestris* causes progressive burial of artifacts by upbuilding over time (T). From Howard et al. (2015)

7.4 Plaggic Anthrosols

Plaggic anthrosols are anthropogenic soils formed by plaggen agriculture, which involved human additions of a mixture of sod and manure over a period of many centuries, primarily occurring in northern Europe during the Middle Ages. This technique was used to improve the productivity of low-fertility sandy soils, and became widespread around A.D. 1000 in an effort to maintain the continuous production of rye. Long-term additions of organic matter led to the development of an artificially overthickened A horizon called a plaggic horizon (IUSS Working Group 2015), or a plaggen epipedon (Soil Survey Staff 2014). Plaggen agriculture continued locally in Europe until the 19th century.

7.4 Plaggic Anthrosols

Table 7.4 Properties of natural soil compared with Hortic Anthosols in Polish convent gardens

Horizon	Depth (cm)	Organic C (%)	Total N (%)	C/N
Natural soil (Haplic Cambisol)				
Ap	0–24	1.50	0.11	13.1
Bw	24–72	0.20	0.02	6.5
C	72–115	0.02	0.02	1.1
2Cg	115–162	0.06	0.02	2.7
Hortic Anthrosol 1 (13th century)				
*Ap	0–35	3.9	0.30	13.1
*A1	35–65	2.2	0.20	14.1
*A2	65–78	2.2	0.20	14.1
*A3	78–122	1.1	0.10	11.5
*A4	122–160	0.8	0.07	10.8
Hortic Anthrosol 2 (17th century)				
*Ap	0–22	2.5	0.2	11.5
*A1	22–93	1.0	0.2	8.9
Hortic Anthrosol 3 (17th century)				
*A1	0–28	3.2	0.2	14.4
*A2	28–49	2.2	0.2	14.4
*A3	49–93	1.1	0.1	11.3

After Gasiorek and Niemyska-Lukaszuk (2008); Horizon designations modified

Table 7.5 Morphological characteristics of a 300–800 year-old agricultural soil in northern Russia

Horizon	Depth (cm)	Description
*A1	0–25	Very dark grayish-brown (10YR3/2) loamy sand; crumb structure; many roots; fragments of peat, brick and charcoal
*A2	25–47	Very dark grayish-brown (10YR3/2) loamy sand; single grain structure; fragments of peat, brick and charcoal
*A3	47–54	Very dark grayish-brown (10YR3/2) loamy sand; single grain structure; 2% charcoal fragments
*A4	54–59	Dark grayish-brown (10YR4/2) silt loam; single grain structure; 5% charcoal fragments
Ab	59–65	Light brownish-gray (10YR6/2) silt loam; single grain structure

Data compiled from Hubbe et al. (2007). Horizon designations modified from original description

Plaggen agriculture was widely practiced in coastal areas of the Netherlands, Germany and Belgium underlain by soils formed in sandy Pleistocene sediments. These areas were often covered in heath, a shrubland habitat characterized by open, low-growing (0.2–2 m tall), woody vegetation found mainly on well drained, infertile, acidic, sandy soils. A major component of heathland in northern Europe is heather (*Calluna vulgaris*), a small, shrubby evergreen ground cover. Plaggen are

Table 7.6 Selected properties of a 300–800 year-old agricultural soil in northern Russia

Horizon	Depth (cm)	pH	Organic C (%)	CEC (cmol kg^{-1})	CaCO$_3$ (%)	Sand (%)	Silt (%)	Clay (%)	Citric-acid P (mg kg^{-1})	Total P (mg kg^{-1})
*A1	0–25	6.5	3.7	11.2	0	77	19	4	759	1344
*A2	25–47	6.8	2.8	9.5	0	75	21	4	791	1412
*A3	47–54	6.7	2.1	7.6	0	75	22	3	583	1159
*A4	54–59	6.8	1.5	5.6	0	73	24	3	456	864
Ab	59–65	6.8	0.5	1.5	0	68	31	1	58	124

Data compiled from Hubbe et al. (2007). Horizon designations modified from original description

Table 7.7 Effects of cultivation on the A horizons of soils at two ancient agricultural sites

Sample type	Bulk density (g cm^{-3})	pH	Organic C (%)	Total N (%)	Total P (mg kg^{-1})	Plant-avail. P (mg kg^{-1})
Sapillo Valley, New Mexico, USA (866 year old)						
Sediment	1.20	5.8	2.1	0.14	444	76
Uncultivated soil	1.30	5.8	1.8	0.13	417	54
Cultivated soil	1.42	6.0	1.0	0.08	343	32
Colca Valley, Peru (466 year old)						
Uncultivated soil	1.38	7.2	1.35	0.12	773	12
Cultivated soil	1.35	6.7	1.58	0.16	1313	44

After Sandor (1993)

Table 7.8 Properties of ancient buried agricultural soils in eastern Grand Canyon area, Arizona, USA (Davis et al. 2000)

Horizon	Depth (cm)	pH	Organic C (%)	EC (dS m^{-1})	Sand (%)	Silt (%)	Clay (%)	CaCO$_3$ (%)	SAR
Cultivated Haplocalcid ("Hortic" Anthrosol, northern Arizona)									
C	0–8	7.5	0.3	0.6	79	20	1	5.7	3.0
*Akb	8–10	7.9	1.0	3.0	51	41	8	8.6	15.0
*Bkb	10–36	8.6	0.4	101.3	27	61	12	14.8	265.0
2*Akb	36–41	9.3	0.5	142.7	31	54	15	13.5	256.0
2*BKb	41–53	9.3	0.4	78.4	70	26	4	15.8	141.0
3*Akb	53–58	8.7	0.9	158.0	54	34	12	13.7	187.0
3C	58–94	8.4	0.3	59.4	54	38	8	21.1	86.0

pieces of sod from heather or grass that were cut, rolled up, mixed with animal droppings, and then applied to fields as an agricultural amendment. It was necessary to obtain as much manure as possible, hence livestock was generally stabled indoors at night in order to concentrate the manure. In the typical scenario, a layer

7.4 Plaggic Anthrosols

of sod ∼3 cm thick would be cut from a 5 to 10 ha area of heathland soil, typically only suitable for grazing livestock, about every 10–15 years, and then applied to a 1 ha field plot near a small village. Thus, the organic-rich layer was concentrated in small plots of arable soil, to the detriment of surrounding scavenged areas. The long-term practice of plaggen agriculture created layers of highly fertile topsoil 40–150 cm thick, in contrast to modern cultivated topsoils which are generally no more than about 30 cm thick. Plaggen applications caused field plots to become raised by a height corresponding to the thickness of the plaggic topsoil, whereas the scalped landscapes became depressed. Thus, the ground surface has often been raised 1 m or more by plaggen applications, with a difference of 2 m or more between raised and depressed landscapes in some areas.

A Plaggic Anthrosol typically consists of a relatively homogeneous plaggic horizon overlying remnants of the sandy natural Inceptisol or Spodosol (Podzol) from which it was formed (Fig. 7.1). The A, E, and perhaps part of the Bw or Bhs, horizons of the buried natural soil are often missing, and were presumably incorporated into the plaggic horizon by plowing. The plaggic horizon, which is known in Germany as an Esch horizon (Eschhorizonte), can be black or brown (Table 7.9). Black plaggen are formed from heather sod, whereas grass sod produces brown

Table 7.9 Morphological characteristics of black and brown Plaggic Anthrosols soils from Netherlands

Horizon	Depth (cm)	Description
Black Plaggic Anthrosol		
^A1	0–45	Black (10YR2/1) fine sand; massive; very humose; contains grains of charcoal and small pieces of burnt loam, with some bleached sand grains
^A2	45–75	Very dark gray (10YR2.5/1) fine sand; massive; very humose; contains grains of charcoal and more bleached sand grains than ^A1
^A3	75–110	Very dark grayish-brown (10YR3.5/1.5) fine loamy sandy; massive; contains grains of charcoal, pieces of burnt loam and bleached sand grains
2A1b	110–135	Very dark grayish-brown (10YR3/2) fine sand; moderately humose; abundant charcoal and bleached grains. Represents buried A1 horizon of weakly developed podzol
2B3b	135+	Light yellowish-brown (10YR6/4) fine sand; very weakly humose. Represents weakly developed B horizon of buried soil
Brown Plaggic Anthrosol		
^A1	0–40	Very dark grayish-brown (10YR3/2) fine loamy sand; massive; moderately humose; some grains of charcoal and conspicuous grains of bleached sand
^A2	40–75	Very dark grayish-brown (10YR2.5/2) fine loamy sand; moderately humose. Some weak gray mottles, scattered grains of charcoal and small iron concretions near base
2C1 g	75–100	Yellowish-brown (10YR5/6) sand; weakly humose with some Fe concentrations (5YR5/8)

Data compiled from Pape (1970). Horizons modified from original description

plaggen. Similarly, Esch horizons derived from sand tend to be black or dark gray, whereas those formed in loess or till are dark brown to yellowish-brown (Pape 1970; Blume and Leinweber 2004). The plaggic horizon often contains artifacts such as charcoal, bone, brick, and pottery, and may contain included chunks of subsoil churned up by digging or tilling. A plaggic horizon normally shows plow marks (Fig. 7.6), at least in its lower part, and it may contain remnants of thin stratified beds of sand that were probably produced by surface runoff and later buried. Also common in plaggic horizons are bleached sand grains, which are thought to have been derived from the E horizon of the natural soil. Map delineations with Plaggic Anthrosols tend to be rectilinear anthropogenic landforms that are higher than adjacent land surfaces by as much or more than the thickness of the plaggic horizon.

The plaggic horizon typically contains ∼2.5–4% organic carbon, has an unlimed pH of 3.5–5.0, a BS <50%, high levels of free Fe (Table 7.10), and higher levels of plant-available nutrients compared with the natural soils from which they were derived. Phosphorous levels are often elevated, typically ≥ 250 mg kg^{-1}

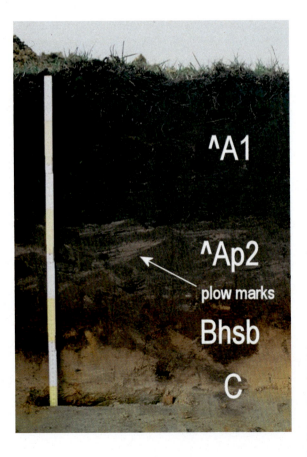

Fig. 7.6 Plaggic Anthrosol with black plaggic horizon about 45 cm thick. Note *plow marks* in layer below (modified from *The Physical Geography of Western Europe* edited by Koster (2005) Fig. 16.14, p.323, used by permission Oxford Univ. Press)

7.4 Plaggic Anthrosols

Table 7.10 Properties of Plaggic Anthrosols from different countries in northern Europe (Pape 1970; Blume and Leinweber 2004; Schnepel et al. 2014)

Horizon	Depth (cm)	pH	Organic C (%)	CEC (cmol kg^{-1})	Total P (mg kg^{-1})	Sand (%)	Silt (%)	Clay (%)	Base sat. (%)	Free Fe (%)
Black Plaggic Anthrosol (Germany)										
^A1p	0–20	4.8	2.9	12.0	650	90	5	5	35	4.5
^A2	20–65	4.0	2.6	12.0	320	88	7	5	8	4.9
2AEb	65–75	4.0	1.6	8.3	170	88	7	5	4	3.7
2Bsb	75–90	4.5	1.0	8.6	200	87	3	10	2	8.0
Black Plaggic Anthrosol (Netherlands)										
^A1	0–45	3.8	3.1	11.3	135	89	6	5	–	0.78
^A2	45–75	3.7	3.6	11.3	117	88	6	6	–	1.04
^A3	75–110	3.9	3.9	11.3	153	86	8	6	–	1.49
2A1b	110–135	4.0	1.7	6.7	113	91	5	4	–	0.57
Black Plaggic Anthrosol (Norway)										
^A1	0–10	5.6	4.0	16.6	2184	68	26	6	58.4	–
^A2	10–80	6.0	2.7	17.0	1669	66	27	67	39.4	–
2Bwb	80–95	6.2	1.7	16.1	1060	67	27	6	24.3	–

Horizons modified from original description

citric acid-soluble P$_2$O$_5$, or ≥1000 mg kg^{-1} total P (Pape 1970; Blume and Leinweber 2004; Schnepel et al. 2014).

Plaggen soils are most widespread in coastal areas around the North Sea in northern Europe where they cover an area of ~6500 km^2 stretching from northeastern Belgium through the Netherlands to northwestern Germany. They are most extensive in the Netherlands where they cover an area of ~4000 km^2, about half of which is characterized by plaggen ≥50 cm thick, and the remainder being 30–50 cm thick. The usual geographic pattern is for sandy soils near villages to have plaggen. Other areas in Europe with plaggen-like soils have been reported in France, Denmark, Ireland, Scotland, northwestern Russia and Norway. The oldest plaggen soils in Europe date to the late Bronze age ~1000 BC, but plaggen soils date primarily from the early Middle Ages (A.D. 500–1000). Plaggen agriculture largely ceased around the beginning of the 20th century as mineral fertilizers became widely available (Pape 1970; Blume and Leinweber 2004; Thomas et al. 2007).

According to the current version of the WRB (IUSS Working Group 2015), Plaggic Anthrosols are characterized by a plaggic horizon ≥50 cm thick. A plaggic horizon has a texture of sand, loamy sand, sandy loam or loam, or a combination of them, contains <20% artifacts, and has a chroma ≤2, a thickness of ≥20 cm, an organic carbon content ≥0.6%, and occurs on locally raised land surfaces. These criteria seem to be intended to restrict plaggic anthrosols to the classic area of plaggen soils in the Netherlands and adjacent parts of Belgium and Germany. Soils in other parts of Europe, which are similar to plaggic anthrosols, are now probably classified mainly as Hortic or Terric Anthrosols. For example, Conry and Diamond (1971) and Conry (1974) thought that all soils on raised land should be classified together as plaggen soils, including soils in Ireland and elsewhere amended with calcareous sands or marl.

Soil Taxonomy (Soil Survey Staff 2014) has a provision for plaggic (plaggen epipedon ≥ 50 cm thick) and haploplaggic (plaggen epipedon ≥ 25 cm, but <50 cm thick) subgroups, but none are yet recognized in the United States. A plaggen epipedon is defined as a thick, human-made mineral surface layer formed in HTM ≥ 50 cm thick, which is found on locally raised landforms and contains artifacts and/or spade marks below a depth of 30 cm, and has colors with a value of ≤ 4 (moist) and a chroma ≤ 2, an organic carbon content $\geq 0.6\%$, and certain moisture and temperature requirements.

7.5 Terric Anthrosols

A Terric Anthrosol is an anthropogenic soil formed by long-term human additions of mainly earth materials (calcareous beach sand, marl, loess, mud, etc.), possibly containing or mixed with organic materials, for the purpose of improving agricultural productivity. Although both terric and plaggic agriculture produced raised land through additions of earthy amendments, the terric cumulic process differs from plaggen method, wherein a subordinate amount of mineral material was added incidentally, in that intentional additions of predominantly mineral material are involved. Examples of Terric Anthrosols include coastal areas in Ireland and England where calcareous beach sand and marl were carted or shipped to areas with nutrient-poor, acid soils in order to improve their suitability for farming. Marl and calcareous sands were used as liming materials in northwestern Germany, Hungary, and ancient Rome. In Mexico, the Aztecs are known to have added organic-rich lacustrine sediments to improve the arability of soils. The Chinese have long used the practice of adding loess, or mud from deltaic-backswamps, as agricultural amendments. Terric Anthrosols are estimated to cover about 1.5 million ha in China.

Gong et al. (1999) referred to Terric Anthrosols as earth-cumulic and mud-cumulic Anthrosols. Earth-cumulic Anthrosols are widespread in the Loess Plateau of the Yellow River valley, where a mixture of loess and organic matter has been used as an agricultural amendment for about 2800 years. Mud-cumulic Anthrosols are common on the deltas of the Zhujiang and Yangtse Rivers. In these areas, farmers excavated mud from backswamp areas and dumped it on nearby areas of higher ground as an ancient form of organic farming. The excavated area becomes a fish pond and the raised areas are used to grow mulberries. The mulberry leaves are used as food by silkworms, whose excrement feeds fish in the pond. Likewise, the fish excrement is used to fertilize the mulberry fields, thus forming a benign biogeochemical cycle. This method dates back at least as far as the Han Dynasty, about 2000 year ago.

The key feature of a Terric Anthrosol is a terric horizon ≥ 50 cm thick, which may overlie an agric horizon (Fig. 7.1). A terric horizon is a mineral surface horizon with all the following properties (IUSS Working Group 2015): (1) a color related to the source material, (2) a BS $\geq 50\%$ (1 M Ammonium acetate, pH 7.0), (3) does not show stratification, (4) occurs locally on raised surfaces and (5) has a thickness of ≥ 20 cm. A terric horizon may contain randomly sorted and distributed stones, and it commonly contains artifacts such as pottery fragments, cultural debris and refuse, that are typically very small (<1 cm in diameter) and much abraded. A terric horizon is usually built up over a long time period, but occasionally is created by single additions of material. Normally the material added is mixed with the original topsoil. Terric soils have a raised surface that may be inferred either from field observation or from historical records. The terric horizon is not homogeneous, but subhorizons are thoroughly mixed. Terric horizons can be somewhat similar to mollic and plaggic horizons.

7.6 Pretic Anthrosols

Pretic Anthrosols are a type of anthropogenic soil characterized by an artificially thickened, organically-enriched *A horizon formed by human additions of biochar and other organic wastes in a tropical climatic setting. Commonly known as Terra Preta de Indio, or Amazonian Dark Earth (Table 7.1), Pretic Anthrosols are found beneath a wide area of tropical rainforest in the Amazon River basin of Brazil and adjacent countries. Pretic Anthrosols were created by unique agricultural activities practiced by pre-Columbian indigenous cultures, and commonly contain artifact assemblages of archaeological interest. The Terra Preta soils produced by these indigenous cultures have agricultural characteristics far superior to those of local natural soils, and these characteristics have persisted for many centuries despite intense weathering in a region where annual precipitation may reach 300 cm per year. This has attracted the interest of many scientists who are studying Terra Preta for their significance in terms of sustainable agriculture and carbon sequestration.

The classic Pretic Anthrosol consists of a black pretic horizon resting on the bright red argillic or oxic subsoil of the Ultisol or Oxisol from which it was formed (Fig. 7.1). Organic matter content may be somewhat elevated in the upper part of the subsoil such that the lower boundary of the pretic horizon is gradual. There is a striking contrast between the black, organic-rich, fertile, pretic horizons, which are 30–200 cm in thickness, and the natural topsoils which are only 10–20 cm thick, light-colored and infertile (Fig. 7.7). Three distinct subhorizons can often be distinguished in a Pretic Anthrosol: (1) a thick, black, sandy topsoil with abundant pottery sherds and other artifacts, (2) a mottled transitional horizon characterized by abundant organic cutans on ped facies and root channels, and (3) a lighter colored zone with higher clay content and fewer cutans (German 2003).

A pretic horizon has the following characteristics according to the WRB (IUSS Working Group 2015): (1) a Munsell color with a value ≤ 4 and a chroma ≤ 3 (moist), (2) $\geq 1\%$ organic carbon content, (3) exchangeable Ca + Mg (1 M ammonium acetate) ≥ 2 cmol kg^{-1}, (4) ≥ 30 mg kg^{-1} extractable P (Mehlich 1),

Fig. 7.7 Profile of Pretic Anthrosol from the central Amazon River basin estimated to be ~1500 years old. Note abundant artifacts, mainly potsherds (from Certini G, Scalenghe R, The Holocene (v. 21, issue 8), pp. 1267–1274, © 2011 by The Holocene; Used by permission SAGE Publications, Ltd.)

(5) ≥1% artifacts, or ≥1% anthropogenic charcoal, or evidence of past human occupation in landscape, (6) evidence of animal activity comprising <25% (by volume) of the horizon, and (7) one or more layers with combined thickness of ≥20 cm. Artifacts are typically fragments of pottery, and stone, bone or shell tools. Pretic horizons do not show the level of animal activity required for hortic and irragric horizons. They are distinguished by a great abundance of visible fragments of charcoal and biochar.

In the field, Amazonian Dark Earth soils vary in color from black (10YR2/1), to dark grayish-brown and brown (Table 7.11), hence two basic types are generally distinguished. The black variety is known as Terra Preta (TP) proper, whereas the brown type is called Terra Mulata (TM). TP is found in small patches, averaging 20 ha, but ranging from <1 ha to 500 ha in size. TP soils are often associated with kitchen middens, oven-mounds and other archaeological features, hence they are thought to represent the habitation sites, trash dumps, and vegetable gardens of Amerindian settlements. Pretic horizons contain a variety of pre-Columbian artifacts, and are thought to have formed in situ by human additions of biochar, ash, fish and mammal bones, turtle shells, animal and human feces, and other domestic wastes. In some areas, patches of TP are bounded by broad swaths of TM soils, which are typically dark grayish-brown and contain few, if any, artifacts, but have

7.6 Pretic Anthrosols

Table 7.11 Morphological characteristics of Terra Mulata soils from the Amazon basin in Colombia

Horizon	Depth (cm)	Description
Terra Mulata soil #1		
*A1	0–7	Very dark gray-brown (10YR3/2) fine sandy loam; structureless; friable; contains fine charcoal fragments and few sherds; many roots
*A2	7–36	Dark reddish-brown (5YR2/2) fine loamy sand; weak fine granular structure; friable; common visible pores; contains few fine charcoal fragments and sherds; common roots
*AB	36–49	Very dark brown (10YR2/2) fine loamy sand; moderate fine subangular blocky structure; friable; common visible pores, few sherds; common roots
Bt1	49–74	Dark brown (7.5YR4/4) sandy clay loam, with abundant strong brown (7.5YR5/8) mottling and incipient iron concretions; moderate fine subangular blocky structure; firm; common visible pores, few sherds; common roots
Bt2	74–106	Strong brown (7.5YR5/6) sandy clay loam, with common to abundant, hard red (10R3/6) iron concretions; strong fine subangular blocky structure; firm; common visible pores, few roots
R	106+	Weathered sandstone
Terra Mulata soil #2		
*A1	0–15	Very dark grayish (10YR3/1) sandy loam; weak fine granular structure; very friable; few fine charcoal fragments and sherds; many roots
*A2	15–90	Very dark grayish-brown (10YR3/2) sandy loam; weak fine subangular blocky structure; friable; few fine charcoal fragments and sherds; few to common roots
*AC	90–120	Very dark grayish-brown (10YR3/2) sandy clay loam, with common weathered rock fragments; weak fine subangular blocky structure; friable; very fine charcoal fragments and few sherds; few roots
R	120+	Weathered igneous/metamorphic rock

Data compiled from Eden et al. (1984). Horizons inferred from original description

elevated levels of organic matter in comparison with surrounding undisturbed rainforest soils. TM soils are thought to represent agricultural fields located outside of former villages, which grade outward into unaltered soils of the surrounding forest where only foraging took place. TM soils appear to have been intentionally amended with human and/or animal manure, green manure, charcoal and ashes, fishing and hunting remains, and calcium from ground-up mollusk shells or root accretions.

Natural soils in the Amazon basin are generally Ultisols and Oxisols, which are strongly acid (pH = 4), and characterized by low levels of plant-available nutrients and high, even phytotoxic, levels of exchangeable Al. In contrast, TP soils are less acid (pH 5.2–6.4), have low levels of exchangeable Al, and high levels of P, N, basic cations, Mn and Zn. TP soils often contain >150–200 mg kg^{-1} plant-available P, compared with 5 mg kg^{-1} in natural soils. TM soils have somewhat lower fertility (Table 7.12), but locally can have 50–150 mg kg^{-1} plant-available P. A pretic horizon may have two to three times the amount of organic matter of

Table 7.12 Properties of natural soils compared with Terra Mulata and Terra Preta soils in Colombia and Brazil

Horizon	Depth (cm)	pH	Organic C (%)	CEC (cmol kg^{-1})	Avail. P (mg kg^{-1})	Sand (%)	Silt (%)	Clay (%)	Base sat. (%)
Terra Mulata soil area #1 (Colombia)									
*A	0–10	4.3	3.8	0.94	71	85	8	7	24.0
Bt2	75–80	4.8	1.6	1.49	17	66	10	24	4.7
Natural topsoils area #1 (Inceptisols, Colombia)									
A1	0–10	4.4	2.0	1.16	12	74	11	15	19.8
A1	0–10	4.2	2.1	1.23	13	78	7	15	18.7
A1	0–10	4.5	4.2	1.71	25	54	22	24	21.0
A1	0–10	4.0	4.4	1.98	25	46	22	32	20.7
Terra Mulata soil area #2 (Colombia)									
*A1	0–10	4.7	2.2	0.91	36	73	16	11	41.8
*A2	75–100	4.5	1.2	0.86	15	61	19	20	33.7
Natural topsoils area #2 (Inceptisols, Colombia)									
A1	0–10	4.0	2.3	1.30	11	57	14	29	23.1
A1	0–10	4.4	2.9	1.38	14	53	19	28	22.5
A1	0–10	4.1	3.0	1.94	24	39	36	25	15.5
A1	0–10	4.0	3.5	2.29	16	44	32	24	13.1
Terra Preta soil (Brazil)									
*A	0–30	5.6	3.2	6.62	68.6	60	16	22	93
*AB	30–58	5.2	1.5	2.59	37.4	52	12	37	72
B1	58–96	5.0	0.6	1.46	41.9	44	11	45	58
B2	96–114	4.8	–	1.36	39.2	35	14	50	39
Natural soil (Oxisol, Brazil)									
A	0–16	4.3	4.5	2.61	3.6	38	18	43	10
AB	16–45	4.3	1.9	1.81	1.7	31	20	50	5
B1	45–88	4.3	1.1	1.76	2.1	34	13	53	8
B2	88–110	4.5	–	0.97	0.7	10	19	71	4

Compiled from Eden et al. (1984) and German (2003). *Note* Brazil P data (P$_2$O$_5$) using Mehlich 1 method; Colombia P data using Bray method

natural soils, and may contain up to 9% organic carbon (partly in the form of pyrogenic black carbon) compared with 0.5% in natural soils. As a result, the CEC of TP soils often ranges from 13 to 25 cmol kg^{-1}, compared 12 or less in natural soils (Lima et al. 2002; Glaser and Birk 2012). Pretic horizons also have higher microbial diversity and species richness than natural soils, but they differ from hortic horizons by having a predominance of fungi over bacteria.

As is the case with many anthropogenic soils, particularly those associated with archaeological sites of human habitation, TP soils often contain visible fragments of what appears in the field to be charcoal, defined here as a form of black carbon (BC) generated as a product of incomplete combustion of wood in an aerobic environment. It has been determined that BC often comprises 30–45% of total soil organic carbon in Pretic Anthrosols, and some patches of TP soils have been found

7.6 Pretic Anthrosols

to contain 70% more BC that natural Amazonian soils (Glaser and Birk 2012). These high levels of BC were difficult to explain by charcoal alone because <2–3% of original pre-burn biomass remains in charcoal. Hence, natural fires produce soil surface residues which are relatively low in BC (Atkinson et al. 2010). This led to the discovery that TP soils are characterized by the presence of biochar (Glaser et al. 2002; Solomon et al. 2007). Biochar is formed by pyrolysis, i.e. the thermal degradation of wood under anaerobic conditions (Lehmann et al. 2011), and is highly recalcitrant to microbial biodecomposition. Approximately 50% of the C in the original source is retained in biochar, hence the BC content of biochar is much higher than that of charcoal. It is now widely accepted that biochar, which may have a mean residence time of $\sim$2000 years in soil, is the cause of the black color of TP soils. Although some BC can be produced naturally by fire and soil fungi, biochar must have been added as an agricultural amendment in pre-Columbian times because natural Amazonian soils are not enriched in BC, and TP soils are not forming today as a result of slash-and-burn agriculture.

Biochar may have come from various sources, including pottery kilns, kitchen ovens, smoldering fires used for food and pottery making, as well as so-called slash-and-char agriculture (Mann 2002). It is also possible that it was produced by charring organic waste in primitive ovens with a low-oxygen supply for use as fuel. Biochar was apparently used widely as an agricultural amendment by pre-Columbian indigenous people. Animal and human wastes were also used, as well as mammal and fish bones, which have been found in TP soil profiles. Native South American farmers still apply biochar to their vegetable gardens today. They make it by charring plant debris and soil in home gardens, by covering burning wood with soil and letting it smolder underground, and by collecting wall material from charcoal kilns. They often mix biochar and manure together and apply it to home gardens. Midden areas are used as home gardens, and home gardens are used as trash dumps, by modern indigenous people in the Amazon. It is also well known that Terra Preta soils are still used for horticulture today, and are highly prized by farmers for their sustainable fertility and production potential (Lehmann et al. 2003). Indigenous people can recognize TP soil by tree species assemblages, which differ from those on natural soils.

Pretic anthrosols may collectively cover 10% or more of the Amazon Basin, mainly in Brazil, but including parts of Peru, Colombia, Bolivia, Venezuela and the Guianas. Radiocarbon dates suggest that they were formed primarily between about 500 BC and AD 1500 (Eden et al. 1984; Glaser and Birk 2012). TP soils are found on variable landscape positions ranging from floodplains to interfluvial uplands, but they are most commonly found along major drainages, on non-floodable land, in strategically advantageous positions. They appear to be coincident with areas of dense settlement located where higher land comes close to, and has good visibility of navigable waterways, with smaller sites being located in interfluvial areas. The total Amerindian population of the Amazon is estimated to have been between 5 and 10 million people around AD 1500. Much of the population is thought to have been wiped out soon thereafter by diseases brought over by Spanish and Portuguese explorers.

TP soils are required to have a pretic horizon ≥ 50 cm in order to be classified as a Pretic Anthrosol. They have many similarities to Hortic Anthrosols, and are partly distinguished from them by an Amazonian geocultural setting. Soils similar to Terra Preta have been recognized in Australia (Downie et al. 2011), and northern Germany (Wiedner et al. 2015) on the basis of high levels of black carbon.

7.7 Irragric Anthrosols

Irragric Anthrosols are found in dry regions where irrigation agriculture was, or has been, practiced for a long time. They are known from Mesopotamia, ancient agricultural areas in China, India and Peru, in certain oases of central Asia and the Middle East, and in the desert areas of Australia and the southwestern United States. They are extensive in China, where they cover about 1.5 million ha, and where the irrigation history locally exceeds 2000 years (Gong et al. 1999). Irragric Anthrosols are perhaps most widespread in central Asia, where they cover more than 10 million ha (Kostyuchenko and Lisitsyna 1976). Irrigation is generally required for agricultural activities in climatic areas receiving less than about 60 cm of annual precipitation. Hence, Irragric Anthrosols are found primarily in arid, semi-arid, and Mediterranean climates, i.e. in regions receiving <25, 25–50, and 35–90 cm of annual precipitation, respectively. Soils receiving <20 cm of annual precipitation generally have high soluble salt contents (Pariente 2001). Hence, irrigation agriculture is often associated with salt-affected soils, defined as soils with excessive levels of salinity and/or sodium content. Saline soils are defined as those with an electrical conductivity >4000 $\mu S\ cm^{-1}$, which marks the level at which excess salinity begins to adversely affect the growth of most plants. Sodic soils, with sodium adsorption ratio (SAR) >13, also adversely affect plant growth because of excessively high pH (>8.5) and poor physical characteristics.

Irrigation generally involves one or more canals, which carry water into an agricultural area, and a series of ditches which carry off any excess surface drainage. All waters used in irrigation contain dissolved salts, and because crop roots normally exclude most of the salts while extracting soil moisture, the salts build up in the soil. Unless these salts are leached, they will accumulate in the root zone and severely restrict crop production. In regions receiving <20 cm of annual precipitation, natural rainfall is insufficient for annual leaching, hence irrigation waters must be applied in excess of crop requirements before harmful salts can be leached from the root zone by percolation. By itself, the practice of over-irrigation is only a temporary remedy, and in the long run actually exacerbates the problem because excessive irrigation waters cause the water table to rise, especially in low-lying river valleys like those where irrigation typically is used. Salt-laden groundwater can move horizontally over underlying impervious geologic layers, and ultimately rise toward the surface in low-lying parts of the landscape. Once the water table rises to within about 1 m or less of the surface, capillary rise will cause groundwater to move upward to replace water lost at the surface by high rates of evaporation

7.7 Irragric Anthrosols

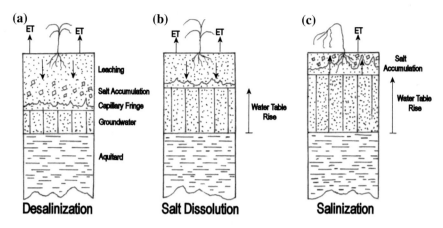

Fig. 7.8 Process of salinization in Irragric Anthrosols: **a** Salts are leached from initially salt-rich surface layer by percolating irrigation water and accumulate in subsoil; **b** Ongoing application of irrigation water causes water table to rise and solubilize salt in subsoil; **c** Capillary fringe approaches ground surface and high rates of evapotranspiration (ET) draw salts into surface layer where reprecipitation takes place

(Fig. 7.8). As the water evaporates, salts accumulate in the rhizosphere, and may eventually reach levels that will restrict the growth of plants. The farmer can no longer leach this salt by over-irrigation because this merely increases the rate of water table rise causing water-logging as well as salinization. Year after year, the salinized zone will creep slowly upslope, and the barren, salt-encrusted, salinized area will become larger and more saline over time. Millions of hectares of land in the United States, Australia, India, the Middle East, the former Soviet Union, and other desert regions have been degraded by this process.

Irragric Anthrosols are natural soils of any type with physical and chemical properties that have been altered by irrigation. The key feature is an irragric horizon (Fig. 7.1), i.e. a mineral surface horizon formed from a pre-existing natural soil by the gradual buildup of fine silt and clay, and possibly fertilizers, soluble salts, organic matter, through continuous application of irrigation water. By definition, an irragric horizon is ≥ 20 cm thick, and has a uniform structure, a higher clay content than the underlying original soil, a carbonate content <20%, an organic carbon content $\geq 0.5\%$, and contains $\geq 25\%$ (by volume) worm burrows, coprolites or other evidence of biological activity (IUSS Working Group 2015). Irragric horizons are known to range from <10 cm, to as much as 10 m in thickness (Gong et al. 1999; Woodson et al. 2015). Irragric Anthrosols are defined as anthropogenic soils with an irragric horizon ≥ 50 cm thick. Soils with an irragric horizon often show evidence of surface raising, and considerable evidence of biological activity. They may have chemical properties which were modified by waterlogging, desalinization or salinization, and also commonly contain artifacts of archaeological significance. An irragric horizon is massive due to continuous plowing, and is therefore readily differentiated from a buried soil, or from stratified irrigation or fluvial sediments, lying beneath it. Some may qualify as umbric or mollic horizons depending on base saturation. Irragric horizons have been documented in irrigation soils of the desert southwestern United States (Davis et al. 2000; Woodson et al. 2015), and in oasis soils of the Karakaum Desert of Turkmenistan (Kostyuchenko and Lisitsyna 1976), the Atacama Desert of Peru (Hesse and Baade 2009), and the Arabian Desert of Oman (Luedeling et al. 2005).

Compared with natural topsoils in the same area, irragric horizons are generally thicker, darker and more clayey, and they have better structure and higher values of pH and bulk density; all attributable to the effects of cultivation. Levels of organic C, N and P may be enriched in ancient cultivated versus uncultivated soils, but these relationships are not always consistent (Homburg et al. 2005), perhaps as a result of changes in organic components following abandonment. Ancient Irragric Anthrosols can retain certain of these properties for thousands of years, and can still be differentiated from natural desert soils, even after many centuries of abandonment (Kostyuchenko and Lisitsyna 1976; Woodson et al. 2015). They can be usually distinguished from natural soils locally by clay or silt content, and may appear on soil maps as long thin "ribbons" which correlate with the traces of ancient canals.

The data in Table 7.13 are for a 560 year-old irrigation soil associated with the Snaketown canal system of the ancient Hohokam culture, who farmed southern Arizona from about AD 450–1450 (Woodson et al. 2015). The irrigation soil has an *A-*Bnkz profile 77 cm thick, lying on the relict argillic horizon of the underlying natural soil. The *A horizon is 35 cm thick, and has an organic carbon content of >0.5% organic carbon (>0.86% organic matter), thus qualifying it as an irragric horizon. However, because the irragric horizon is <50 cm thick, the soil does not technically qualify as an Irragric Anthrosol. Natural soils in the area are saline-sodic Haplocambids and Argids, but the irragric horizon is non-sodic and non-saline, based on an EC <4000 µS cm^{-1}, and a SAR <13. Ancient irragric horizons are often less saline than associated natural topsoils (Kostyuchenko and Lisitsyna 1976; Nordt et al. 2004; Woodson et al. 2015), but the opposite situation has also been documented (Table 7.8). Salinity relationships in an irrigation agricultural setting can have a large amount of geospatial variability (Table 7.13) as a result of differences in microrelief (Kostyuchenko and Lisitsyna 1976), and in the composition of irrigation water (Luedeling et al. 2005).

7.8 Hydraquic Anthrosols

Hydraquic anthrosols are anthropogenic soils formed by wet cultivation in paddies. Paddies are constructed wetlands typically used to produce rice, which is a semi-aquatic food crop that can grow in waterlogged anaerobic soil which is submerged in shallow water for prolonged periods of time. Rice is one of the few crops that can tolerate the excessive wetness characteristic of the moonsonal climate of southeastern Asia. Paddy rice cultivation has been practiced in China for ~7000 year, and is now carried out throughout southeastern Asia, the Philippines,

7.8 Hydraquic Anthrosols

Table 7.13 Properties of a 560 year-old Irragric Anthrosol associated with the ancient Hohokam culture at the Snaketown site in southern Arizona, USA, compared with irrigated soils in Peru and Oman (Nordt et al. 2004; Luedeling et al. 2005; Woodson et al. 2015)

Horizon	Depth (cm)	pH	Organic C (%)	EC (dS m^{-1})	HCO$_3$-ext. P (mg kg^{-1})	Sand (%)	Silt (%)	Clay (%)	CaCO$_3$ (%)	SAR
"Irragric Anthrosol" (Siltic, Sodic; central Arizona)										
*A1	0–4	8.0	0.9	0.7	16	33	60	7	3.91	0.3
*A2	4–16	8.5	0.7	0.9	9	18	69	13	6.58	0.8
*Bk	16–35	8.4	0.5	2.9	7	17	69	14	7.68	5.6
*Bnkz1	35–61	7.9	0.5	14.1	4	18	69	13	6.75	12.8
*Bnkz2	61–77	7.6	0.3	14.1	2	5	89	6	6.43	7.8
2Btnkzb1	77–97	7.5	0.3	22.2	3	3	73	25	6.82	7.3
2Btnkzb2	97–118	7.4	0.3	27.0	2	4	52	44	3.20	6.7
2Btnkzb3	118–144	7.5	0.2	22.2	2	26	55	19	4.06	7.0
2BCnkzb	144–177+	7.6	0.1	15.5	1	20	70	10	6.60	7.2
"Irragric Anthrosol" (Peru)										
*A	0–30	8.0	0.5	13.0	406	88	8	4	–	4.3
*Bw	30–60	7.1	0.1	12.5	–	52	32	16	–	0
"Irragric Anthrosol" (Oman)										
*A1	0–15	8.3	3.7	1.40	–	–	–	–	37.9	–
*A2	15–45	8.4	3.0	0.88	–	–	–	–	38.9	–
*A3	45–2100	8.4	1.6	4.83	–	–	–	–	43.4	–

Horizons modified from original description

and elsewhere including parts of Italy, France, Haiti and the United States. Hydragric soils are estimated to cover about 25–30 million ha in China (Gong et al. 1999) and, collectively, perhaps 155 million ha of Earth's surface (Kogel-Knabner et al. 2010).

Ancient methods of cultivating and harvesting rice using human and animal labor are still practiced in most Asian countries. Paddy fields may be situated in naturally low-lying wetland areas such as floodplains, but they are also commonly constructed by terracing hill slopes. The fields are first leveled by dragging a log over them, and then a system of small earthen berms or dikes are constructed to impound river or rain water. Paddy fields are prepared (puddled) by plowing or harrowing the water-saturated soil often using water buffalo, which are naturally adapted for life in wetlands. Repeated tillage at the same depth, year after year, creates an impermeable, compacted layer (plow pan or plow sole) just below the depth of disturbance by the plow. The plowed soil is fertilized (using dung or sewage), and then inundated by a water layer that is usually between 5 and 15 cm deep, although in areas of heavy rains rice may be grown in water 30–120 cm (1–4 feet) deep. Rice seedlings are germinated in seedling beds, grown for 30–50 days, and then transplanted by hand to the paddy fields, where irrigation is maintained during the growing season (4–5 months) by a system of dikes and canals. Paddy fields are drained and dried at harvest, and subsequently re-flooded and replanted after an interval of time ranging from a few weeks to as long as 8 months.

Fig. 7.9 Profile of Hydragric Anthrosol from northwestern Italy ~500 years old. Anthraquic horizon (0–55 cm depth), comprised of puddled layer and plow pan, overlies hydragric horizon (~55–80 cm depth) (from Certini G, Scalenghe R, The Holocene (v. 21, issue 8), pp. 1267–1274, © 2011 by The Holocene; Used by permission SAGE Publications, Ltd.)

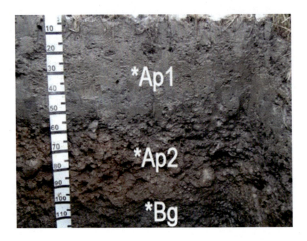

The long-term additions of organic matter, the effects of plowing, and the repeated human-induced fluctuations from anaerobic to aerobic conditions lead to the development of unique morphological features in paddy soils. They vary widely depending on the properties of the natural soils from which they were formed, and local water table conditions. However, paddy soils generally consist of an anthraquic horizon overlying a hydraquic horizon (Fig. 7.1). The cultivated plow layer (Ap1) and a plowpan (Ap2) make up the anthraquic horizon, and a more or less gleyed illuvial subsoil (Bg or Btg horizon) comprises the hydragric horizon (Fig. 7.9). A bleached eluvial layer (Eg horizon) or albic horizon is sometimes located below the plowpan (Tan 1968; Gong 1983; Chen et al. 2011). The Ap1 is enriched in organic matter as a result of human additions of manure, and the Ap2 is typically densic with platy structure. Redoximorphic features are especially characteristic of Hydraquic Anthrosols because of repeated wetting and drying.

When the paddy soil is submerged, the Ap1 and Ap2 undergo reduction. If the water table is low enough, exchangeable bases, Fe, Mn, Si and P are solubilized, translocated and precipitated in the subsoil below the Ap2. When reduced Fe and Mn reach the oxygenated surfaces of the rice roots, or the oxidized zone below the plow pan, they form various types of redoximorphic accumulations. Hence, the rhizosphere of the rice is often characterized by reddish-brown Fe-oxide accumulations along root channels. When the land is drained during harvest, almost the entire profile above the water table is re-oxidized, giving it a highly mottled appearance. Over time, Fe and Mn are permanently lost from the topsoil. The eluviated Fe and Mn, along with some phosphate, accumulate below the plow zone producing a B horizon that is enriched in Fe and Mn. Paddy soils are distinguished by this unique process of reductive eluviation and oxidative illuviation (Ponnamperuma 1972), along with melanization of the plow layer driven by human additions of manure and other forms of organic matter. The leaching of Fe and Mn increases over time, possibly forming a bleached layer, i.e. an Eg horizon (Table 7.14), or tongues of albic material (Bartelli 1973).

7.8 Hydraquic Anthrosols

Table 7.14 Morphological characteristics of 50 year-old and 1000 year-old paddy soils in China

Horizon	Depth (cm)	Description
50 year-old paddy soil (Hydragric Anthrosol)		
*Ap1	0–16	Grayish-brown (10YR5/2) silt loam, with few fine prominent red (2.5YR5/8) masses; weak fine granular structure; loose; many fine roots; clear smooth boundary
*Ap2	16–25	Yellowish-brown (10YR5/4) silt loam; cloddy structure; firm; few fine roots; clear smooth boundary
*Bg1	25–50	Yellowish-brown (10YR5/4) silt loam, with few fine prominent brownish-yellow (10YR6/6) masses; weak fine subangular blocky structure; friable; gradual smooth boundary
*Bg2	50–70	Pale brown (10YR6/3) silt loam, with few fine prominent brownish-yellow (10YR6/6) masses; weak fine subangular blocky structure; friable; gradual smooth boundary
*Bg3	70–100	Grayish-brown (10YR5/2) silt loam, with few fine prominent brownish-yellow (10YR6/6) masses; weak fine subangular blocky structure; firm; gradual smooth boundary
BCg	100–120	Pale brown (10YR6/3) silt loam, with common fine prominent brownish-yellow (10YR6/6) masses; weak fine subangular blocky structure; friable
1000 year-old paddy soil (Hydragric Anthrosol)		
*Ap1	0–16	Dark gray (10YR4/1) silt loam, with common fine prominent brownish-yellow (10YR 6/8) masses; moderate medium granular structure; loose; many fine roots; clear smooth boundary
*Ap2	16–25	Gray (5Y5/1) silty clay loam, with common medium faint gray (5Y5/1) depletions; cloddy structure; very firm; few fine roots; clear smooth boundary
*Btg1	25–50	Gray (5Y5/1) to brownish-yellow (10YR6/6) silty clay loam, with many coarse prominent brownish-yellow (10YR6/6/) masses, and common medium distinct light gray (5Y7/1) depletions; strong coarse subangular blocky structure; firm; gradual smooth boundary
*Btg2	50–70	Gray (5Y5/1) silty clay loam, with many coarse prominent reddish-yellow (5YR6/6) masses, and common medium distinct light gray (5Y7/1) depletions; strong coarse subangular blocky structure; firm; gradual smooth boundary
*Btg3	70–85	Gray (5YR5/1) silty clay loam, with many coarse prominent reddish-yellow (7.5YR6/6) masses, and common medium distinct light gray (5Y7/1) depletions; strong coarse subangular blocky structure; firm; abrupt wavy boundary
Ab	85–100	Very dark gray (10YR3/1) silty clay loam; moderate medium subangular blocky structure; firm; many medium, fine and very fine roots; neutral (pH 7.3); abrupt wavy boundary
Bb	100–120	Gray (5Y5/1) silty clay loam, with many coarse distinct yellow (10YR7/8) masses; strong coarse subangular blocky structure; firm

Data compiled from Chen et al. (2011). Horizons modified from original description

Chen et al. (2011) studied a chronosequence of ancient paddy soils ranging in age from 50 to 1000 years. The Ap1 horizons were dark and organic-rich, and Ap2 horizons were densic with platy structure. Soils <300 years old had a Bg horizon which was yellowish-brown with no redox depletions, whereas soils 700–1000 years old had a Btg horizon with a predominantly gray soil matrix characterized by prominent redox depletions. Soils 300–700 years old were leached of carbonate, and an accumulation of illuvial clay was seen in soils 700–1000 years old. Fe and Mn were depleted in Ap and Eg horizons, whereas Bg and Btg horizons were enriched in Fe and Mn. The soils showed a decrease in pH over first 300 years from an initial value of about 8.1, then pH stabilized at ~6.4. Carbonate content decreased over time, and was completely leached during a span of 300–700 years. This is consistent with the fact that calcite is unstable under reducing conditions. Organic matter content initially increased, but reached equilibrium in only about 50 years. Wet cultivation favors the accumulation of organic matter, hence paddy soils are generally enriched in organic carbon, compared to native soils, with fulvic acid predominating over humic acid (Gong 1983). If the water table is high, some paddy soils become calcareous with free $CaCO_3$ accumulating in the puddled layer and plow pan as a result of over-liming or irrigation with Ca-rich water. Levels of plant-available P can be high in younger paddy soils, perhaps as a result of its release during Fe-oxide dissolution (Gong 1983), and then decrease over time (Chen et al. 2011).

The data in Table 7.15 are from a study of paddy soils in Bangladesh (Brinkman 1977). The results are similar to those of Chen et al. (2011), in that, the paddy soil is enriched in organic carbon relative to the parent from which it was formed. However, the CEC of the paddy soil is distinctly lower than that of the parent soil, and the value of free Fe in the surface horizon is also much lower than that of the underlying horizons. These results were thought to indicate that Fe^{2+} was released to solution by protonation, and then adsorbed onto the exchange sites of clay minerals and organic matter. Consequently, basic cations were desorbed and underwent hydrolysis, thus generating OH^-. During each subsequent drying event, sorbed Fe^{2+} was oxidized to form ferrihydrite and H^+. This cyclic process was

Table 7.15 Properties of a paddy soil from Bangladesh (Brinkman 1977)

Horizon	Depth (cm)	pH	Organic C (%)	CEC (cmol kg^{-1})	P$_2$O$_5$ (mg kg^{-1})	Sand (%)	Silt (%)	Clay (%)	Free Fe (Fe$_2$O$_3$) (%)	Soluble salts (cmol kg^{-1})
*Apg1	0–8	4.9	0.6	4.0	200	5	83	12	0.89	0.05
*Apg2	8–13	5.0	0.4	6.3	200	2	73	25	1.66	0.08
*Bg1	13–18	5.0	0.5	7.9	200	2	70	28	1.54	0.10
*Bg2	18–30	4.9	0.2	6.6	200	1	69	30	2.46	0.09
*Bg3	30–41	4.9	0.1	8.7	200	2	65	33	2.31	0.09
BCg	41–58	5.1	bdl	11.4	200	3	53	42	1.69	0.12
Cg1	58–97	5.3	bdl	13.8	200	3	55	42	1.84	0.19
Cg2	97–127	5.6	bdl	15.6	100	3	54	43	1.50	0.23
Cg3	127–152	5.8	0.2	17.3	100	1	55	44	1.55	0.05

Horizons modified from original description

referred to as ferrolysis by Brinkman (1970, 1977) who thought that excess H^+ generated by this process could eventually cause dissolution of clay, and a reduction of cation exchange capacity. However, this change in clay mineralogy possibly only affects K-bearing clay minerals, and ferrolysis apparently occurs only under certain circumstances (Eaqub and Blume 1982; Chen et al. 2011).

Paddy soils are formed in human-altered material and classified into Anthraquic subgroups, according to Soil Taxonomy (Soil Survey Staff 2014). Anthraquic soils are characterized by anthric saturation, i.e. the soil undergoes continuous or periodic human-induced saturation and chemical reduction, as indicated by redoximorphic features. Anthraquic soils must have both: (1) a tilled surface layer and a directly underlying slowly permeable layer (plow pan) that for three months or more is saturated and reduced, and has a matrix chroma of ≤ 2, and (2) a subsurface horizon characterized by redoximorphic features in the form of depletions and/or concentrations. Paddy soils in the 1000 year-old chronosequence studied by Chen et al. (2011) were found to grade from Entisols to Typic Haplanthrepts, and then Anthraquic Hapludalfs, with increasing soil age.

Paddy soils are classified as Hydragric Anthrosols according to the WRB (2015). Hydragric soils have an anthraquic horizon and an underlying hydragric horizon within 100 cm of the surface. An anthraquic horizon is defined as an anthropogenic surface horizon ≥ 20 cm thick comprised of puddled layer (Ap1), and a plow pan (Ap2), characterized by redoximorphic features. The puddled layer has a Munsell hue of 7.5YR or yellower, a value ≤ 4, and a chroma of ≤ 2, and has sorted soil aggregates and vesicular pores. The plow pan is typically densic with platy structure, and characterized by redoximorphic features in the form of Fe–Mn depletions (mottles) or concentrations (coatings). A hydragric horizon (Bg or Btg) is a human-induced subsurface horizon that occurs below the plowpan, and has either reduction features in pores, or segregations of Fe and/or Mn in the matrix, as a result of an oxidative soil environment. It usually shows gray clay-fine silt and clay-silt-humus cutans on ped faces.

References

Atkinson CJ, Fitzgerald JD, Hipps NA (2010) Potential mechanisms for achieving agricultural benefits from biochar application to temperate soils: a review. Plant Soil 337:1–18

Bartelli LJ (1973) Soil development in loess in the southern Mississippi valley. Soil Sci 115:254–260

Blume HP, Leinweber P (2004) Plaggen soils: landscape history, properties, and classification. J Plant Nutr Soil Sci 167:319–327

Brevik EC, Hartemink AE (2010) Early soil knowledge and the birth and development of soil science. Catena 83:23–33

Brinkman R (1970) Ferrolysis, a hydromorphic soil forming process. Geoderma 3:199–206

Brinkman R (1977) Surface-water gley soils in Bangladesh: genesis. Geoderma 17:111–144

Certini G, Scalenghe R (2011) Anthropogenic soils are the golden spikes for the Anthropocene. The Holocene 21: 1267–1274

Chen LM, Zhang GL, Effland WR (2011) Soil characteristic response times and pedogenic thresholds during the 1000-year evolution of a paddy soil chronosequence. Soil Sci Soc Am J 75:1807–1820

Conry MJ (1974) Plaggen soils—a review of man-made raised soils. Soils Fertilizers 37:319–326

Conry MJ, Diamond JJ (1971) Proposed classification of plaggen soils. Pedologie 21:152–161

Darwin C (1881) The formation of vegetable mould through the action of worms with observations on their habits. John Murray, London

Davis SW, Davis ME, Lucchitta I, Finkel R, Caffee M (2000) Early agriculture in the eastern Grand Canyon of Arizona, USA. Geoarchaeology 15:783–798

Downie AE, Van Zwieten L, Smernik RJ, Morris S, Munroe PR (2011) *Terra Preta Australis*: Reassessing the carbon storage capacity of temperate soils. Aric Ecosyst Environ 140:137–147

Eaqub M, Blume HP (1982) Genesis of a so-called ferrolyzed soil of Bangladesh. Z Pflanz Bodenk 145:470–482

Eden MJ, Bray W, Herrera L, McEwan C (1984) *Terra Preta* soils and their archaeological context in the Caqueta basin of southeast Colombia. Am Antiq 49:125–140

Gasiorek M, Niemyska-Lukaszuk J (2008) Resources and fractional composition of humus in soils of convent gardens of Cracow. Polish J Soil Sci 41:1–11

German LA (2003) Historical contingencies in the coevolution of environment and livelihood: contributions to the debate on Amazonian Black Earth. Geoderma 111:307–331

Glaser B, Lehmann J, Zech W (2002) Ameliorating physical and chemical properties of highly weathered soils in the tropics with charcoal - a review. Biol Fert Soils 35: 219–230

Glaser B, Birk JJ (2012) State of the scientific knowledge on properties and genesis of Anthropogenic Dark earths in central Amazonia. Geochim et Cosmo Acta 82:39–51

Gong Z (1983) Pedogenesis of paddy soil and its significance in soil classification. Soil Sci 135:5–10

Gong Z, Zhang X, Luo G, Shen H, Spaargaren O (1997) Extractable phosphorous in soils with a fimic epipdeon. Geoderma 75:289–296

Gong Z, Zhang G, Luo G (1999) Diversity of anthrosols in China. Pedosphere 9:193–204

Hesse R, Baade J (2009) Irrigation agriculture and the sedimentary record in the Palpa Valley, southern Peru. Catena 77:119–129

Homburg JA, Sandor JA, Norton JB (2005) Anthropogenic influences in Zuni agricultural soils. Geoarchaeology 20:661–693

Howard JL, Dubay BR, Daniels WL (2013) Artifact weathering, anthropogenic microparticles, and lead contamination in urban soils at former demolition sites, Detroit. Michigan: Environ Pollution 179:1–12

Howard JL, Ryzewski K, Dubay BR, Killion TK (2015) Artifact preservation and post-depositional site-formation processes in an urban setting: a geoarchaeological study of a 19th century neighborhood in Detroit. Michigan J Archaeol Sci 53:178–189

Hubbe A, Chertov O, Kalinina O, Nadoporozhskaya M, Tolksdorf-Lienemann E, Giani L (2007) Evidence of plaggen soils in European north Russia (Arkhangelsk region). J Plant Nutr Soil Sci 170:329–334

IUSS Working Group WRB (2015) World Reference Base for Soil Resources 2014, update 2015 International soil classification system for naming soils and creating legends for soil maps World Soil Resources Reports No. 106. FAO, Rome

Kogel-Knabner I, Amelung W, Cao Z, Fiedler S, Frenzel P, Jahn R, Kalbitz K, Kolbl A, Scholter M (2010) Biogeochemistry of paddy soils. Geoderma 157:1–14

Koster EA (2005) The physical geography of western Europe. Oxford University Press, 472 p

Kostyuchenko VP, Lisitsyna GN (1976) Genetic characteristics of ancient irrigated soils. Soviet Soil Sci 8:9–18

Lehmann J, Da Silva Jr JP, Steiner C, Nehls T, Zech W, Glaser B (2003) Nutrient availability and leaching in an archaeological Anthrosol and a Ferrasol of the central Amazon basin: fertilizer, manure and charcoal amendments Plant Soil 249: 343–357

Lehmann J, Rillig MC, Thies J, Masiello CA, Hockaday WC, Crowley D (2011) Biochar effects on soil biota—a review. Soil Biol Biochem 43:1812–1836

Lima HN, Schaefer CER, Mello JWV, Gilkes RJ, Ker JC (2002) Pedogenesis and pre-Columbian land use of "Terra Preta Anthrosols" (Indian black earth") of western Amazonia. Geoderma 110:1–17

Luedeling E, Nagieb M, Brandt M, Deurer M, Buerkert A (2005) Drainage, salt leaching and physic-chemical properties of irrigated man-made soils in a mountain oasis of northern Oman. Geoderma 125:273–285

Mann CC (2002) The real dirt on rain forest fertility. Science 297:920–923

Nordt L, Hayashida F, Hallmark T, Crawford C (2004) Late Prehistoric soil fertility, irrigation management, and agricultural production in northwest coastal Peru. Geoarchaeology 19:21–46

Pape JC (1970) Plaggen soils in the Netherlands. Geoderma 4:229–255

Pariente S (2001) Soluble salt dynamics in the soil under different climatic conditions. Catena 43:307–321

Ponnamperuma FN (1972) The chemistry of submerged soils. Adv Agron 24:29–96

Sandor JA (1993) Long-term effects of prehistoric agriculture on soils: examples from New Mexico and Peru. In: Holliday VT (ed) Soils in archaeology—Landscape evolution and human occupation. Smithsonian, Washington, D.C., pp 217–245

Schnepel C, Potthoff K, Elter S, Giani L (2014) Eviednce of plaggen soils in SW Norway. J Plant Nutr Soil Sci 177:638–645

Smith G (1986) The guy Smith interviews: rationale for concepts in soil taxonomy. In: Forbes TR (ed) USDA Soil Management Service Technology Monog. 11, Cornell University, Ithaca, NY

Soil Survey Soil Staff (1975) Soil taxonomy—a basic system of soil classification for making and interpreting soil surveys. U. S. Dept. Agric., Agricultural handbook 436, Washington, DC

Soil Survey Staff (2014) Keys to soil taxonomy, 12th edn. U.S. Department of Agriculture, Natural Resources Conservation Service, 372 p

Solomon D, Lehmann J, Thies J, Schafer T, Liang B, Kinyangi J, Neves E, Peterson J, Luizao F, Skjemstad J (2007) Molecular signiature and sources of biochemical recalcitrance of organic C in Amazonian Dark Earths. Geochim Cosmochim Acta 71:2285–2298

Tan KH (1968) The genesis and characteristics of paddy soils in Indonesia. Soil Sci. Plant Nut. 14:117–121

Thomas J, Simpson I, Davidson D (2007) GIS mapping of anthropogenic soils in Scotland: Investigating the location and vulnerability of Scottish plaggen-type soils. Atti Soc Tosc Sci Nat MemSeries A 112:85–90

Tisdale SL, Nelson WL, Beaton JD (1985) Soil fertility and fertilizers. Macmillan, New York 754 pp

Wiedner K, Schneeweib J, Dippold MA, Glaser B (2015) Anthropogenic Dark Earth in northern Germany – The Nordic analogue to *terra preta de indio* in Amazonia. Catena 132:114–125

Woodson MK, Sandor JA, Strawhacker C, Miles WD (2015) Hohokam canal irrigation and the formation of Irragric Anthrosols in the middle Gila River valley, Arizona, USA. Geoarchaeology 30:271–290

Zhang M, Ma L, Wenqing L, Chen B, Jia J (2003) Genetic characteristics and taxonomic classification of Fimic Anthrosols in China. Geoderma 115:31–44

Chapter 8
Anthropogenic Soils in Archaeological Settings

In addition to the Plaggic and Pretic Anthrosols discussed in Chap. 7, four other types of anthropogenic soil can be distinguished at archaeological sites: (1) Midden, (2) European Dark Earth, (3) Cemetery, and (4) Burial mound. Midden soils are classified as Hortic Anthrosols whereas burial mound soils are Spolic Technosols. Cemetery soils are Anthrosols, but they require a unique qualifier which has yet to be defined in the WRB. European Dark Earth refers to a type of buried soil (paleosol). All contain artifacts and are often black as a result of elevated levels of humin or black carbon (charcoal). Kitchen midden and midden-mound soils characterize archaeological sites where organic (food) wastes from human occupation were deliberately dumped. They have P-rich topsoils which were the inspiration for the anthropic epipedon as originally defined in U.S. Soil Taxonomy. Shell-midden and other types of midden soils may or may not be P-rich. Midden soils are typically the overthickened *A horizon of a natural soil, and often have elevated levels of pH, exchangeable bases, and carbonate content such that they are classified in a different soil order from that of local natural soils. European Dark Earth paleosols probably represent the buried *A horizon of one or more soils, found beneath the oldest parts of London, England, and other European cities dating from the late Roman or Medieval periods (5th to late 11th century). European Dark Earth resembles a Hortic Anthrosol, and was formed partly at archaeological sites of human occupation, and partly as the result of agricultural (plaggen?) activities. Grave soils are distinguished by a unique type of subsurface horizon formed by the residual accumulation of organic matter derived from the decomposition of a corpse, accompanied by an artifact assemblage comprised of grave goods. This ("necric") horizon has not yet been defined in the WRB, hence classification of grave soils remains problematic. Studies of burial mound soils in North America showed that cambic horizons were well expressed in anthropogenic soils ~500–2000 years old, proto-argillic horizons were evident after ~2000–2500 years of pedogenesis, and argillic horizons were formed in approximately 2500–3000 years.

8.1 Introduction

The presence of pedological features indicative of human activity, especially in association with artifacts or debitage, is considered diagnostic for the recognition of an archaeological site. Hence, anthropogenic soils are fundamentally important to the archaeologist. Anthropogenic soils are characteristic of archaeological sites of human habitation and where humans have impacted soils through agricultural, ceremonial, and burial activities. Sites can range from a surface soil containing simple stone tools at some remote rural location, to complex sequences of artifact-rich prehistoric or historic archaeological strata in a densely populated urban setting. Anthropogenic soil features are used to interpret archaeological context and relative age relationships, and to infer past human behaviors. Anthropogenic soils are helpful for defining site boundaries, identifying activity areas and features, defining stratigraphic relationships, and aiding in functional interpretations (Limbrey 1975; Holliday 2004). They also may contain extensive paleoenvironmental information in the form of phytoliths, plant pollen, etc. Because the identification of human impacts on soils is so critical for distinguishing between natural and artificial soil features, anthropogenic soils have been studied by archaeologists perhaps more than any other type of scientist.

Archaeological soils are often natural soils that have been changed drastically by long-term human additions of organic (e.g., plant materials, soil organic matter, charcoal, soot) and/or inorganic (e.g., pottery, brick, bone, mollusk shells, ash) materials, perhaps taking place over many centuries. They also may have been altered significantly by physical disturbance such as human-induced compaction by trampling, excavation and deposition. Thus, archaeological sites may be characterized by anthropogenic soils formed in human-altered material (e.g., middens, agricultural soils, grave soils), or in human-transported material (e.g., burial mound soils).

Six basic types of archaeological soils can be distinguished: (1) Plaggic, (2) Pretic, (3) Midden, (4) European Dark Earth, (5) Cemetery, and (6) Burial mound. These soils vary widely in their physical and chemical properties, but there are two characteristics which they tend to have in common: (1) enrichment in organic C and certain other chemical elements, especially phosphorous, and (2) the presence of artifacts or biofacts. Biofacts are biological artifacts, i.e. biological materials which are present at an archaeological site as a result of human handling (e.g., used for food or medicine), but which have not been manipulated or altered significantly by humans (e.g., for use as a tool). Biofacts include antler, bone, shells, leaves, seeds, and wood. They are indicators of human occupation, and if they provide information about how past human societies interacted with their environment they may be referred to as ecofacts. Plaggic, Pretic, midden and European Dark Earth soils are all organic-rich as a result of the disposal, or purposeful additions, of organic wastes, whereas burial mound and cemetery soils may or may not be organic-rich, depending on the nature and circumstances of human interment and decomposition. European Dark Earth soils differ from Plaggic and

8.1 Introduction

Pretic types in being associated with an urban setting, whereas midden, burial mound and cemetery soils may or may not be located in an urban setting. European Dark Earth refers to buried soils (paleosols), hence they differ from Amazonian Dark Earth (Pretic Anthrosol), which is a surface soil. Plaggic and Pretic Anthrosols were discussed in Chap. 7, and their further archaeological implications are beyond the scope of this book. Hence, this chapter will focus on midden, European Dark Earth, cemetery, and burial mound soils.

8.2 Midden Soils

A midden (kitchen midden) is an archaeological site where the wastes from human occupation were deliberately dumped. The "classic" kitchen midden is located in a non-urban, often remote setting where prehistoric nomadic people of North and South America, and the early inhabitants of Europe, settled temporarily at a particular site, perhaps returning to the same spot every year, and discarded food waste onto the soil. Over many decades or centuries, the midden developed a thick, black, organic-rich topsoil containing animal bones, mollusk shells, seeds and other plant materials, along with charcoal, ash and perhaps miscellaneous pottery sherds, stone tools, etc. Such middens are common in coastal marine settings, but also can be found associated with bodies of freshwater, including the playas of dry Pleistocene lakes in modern desert settings. The classic kitchen midden was the inspiration for the anthropic epipedon of Soil Taxonomy (Smith 1986), which was originally defined as similar to a mollic epipedon, but containing elevated levels of citric acid-extractable phosphorous. The term midden is now used for any site where archaeological materials related to human habitation have been dumped. Middens are often distinguished according to their primary feature (e.g., shell-midden, bone-midden, ash-midden, sheet-midden, burnt rock-midden, etc.). Kitchen middens found in an urban context (Fig. 8.1) are sometimes referred to generically as Dark Earth. Middens are highly valued by archaeologists because they not only indicate sites of human occupation, but often provide a wealth of information about human behavior, living activities, and local ecology.

Middens are generally localized sites, ranging from <0.5 to several hectares in size, where domestic wastes, typically associated with food preparation, from one or more households were deliberately dumped. The size of a midden is a function of population size, and the length of time the site was active. A midden may be in the form of a pit, a mound, or a layer in a stratigraphic sequence. A kitchen midden usually forms as a result of repeated episodes of dumping, but may be created by a single ceremonial feast. Its composition depends on local resources, and the societal level of technological development. Middens commonly contain ash, charcoal, tools, ceramics, slag, construction materials, seeds, shells, and animal bone. Occasionally human burials are present.

Shell-middens are a common type of midden known worldwide from marine and lacustrine shoreline sites. Shell-middens are usually in the form of mounds (locally

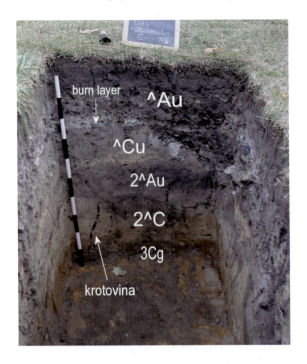

Fig. 8.1 Kitchen midden at an historic (late 19th to early 20th century) urban archaeological site (Detroit, Michigan USA). Thick ^Au horizon contains an ash layer, and abundant artifacts indicative of human occupation (brick, mortar, nails, glass, bones, pottery sherds, coins, etc.). Abundant krotovina in upper C horizon are burrows of the earthworm *Lumbricus terrestris*. Scale in 10 cm intervals. Photo by Krysta Ryzewski

8–9 m or more high) composed mainly of a mixture of soil and mollusk shells (mainly pelecypods), but they can also contain fish bones, sea urchin spines, the remains of other marine animals, bird and mammal bones, ash deposits, and artifacts made from stone, bone, antler or shell (Sawbridge and Bell 1972). Coastal shell-middens are usually located immediately shoreward of a beach, and possibly along bays and tidal channels. Shell-midden soils often show little horizonation, even if very old, and are usually comprised simply of an *A horizon, resting on native soil or bedrock. The lack of morphological development can be attributed to the fact that midden materials, such as shell and bone, are a poor source of pedogenic clay and Fe-oxide. The record of anthropogenic shell-middens stretches back at least 165,000 years, but such features became especially widespread on a global scale between about 12,000 and 6000 years ago (Erlandson 2013). Ash-middens are composed of hearth layers, often mixed with bone, coprolites and local sediments, which are commonly phytolith-rich. A sheet-midden is formed where scraps of food waste are thrown out for chickens and other farm animals to eat. Burnt rock-middens, comprised of piles of fire-cracked rock in the form of mounds or domes, are found in western and central Texas, USA where they are 300–5000 years old (Holliday 2004).

Kaufman and James (1991) studied some shell-midden soils in the Chesapeake Bay region of the U.S. that were estimated to be between ∼300 and 1000 years old. Data for a representative profile are given in Table 8.1. The soils had *A horizons up to 63 cm thick that were very dark gray to black, and contained mainly

8.2 Midden Soils

Table 8.1 Distribution of CaCO₃, organic matter, and citric acid-extractable P as a function of depth in an oyster shell-midden soil, Chesapeake Bay area, USA (Kaufman and James 1991)

Horizon	Depth (cm)	Color	CaCO$_3$ (%)	Organic C (%)	P$_2$O$_5$ (mg kg^{-1})
*A1	0–30	10YR2/2	29.7	3.5	1049.4
*A2	30–63	10YR2/1	18.8	3.8	1326.1
2BA	63–93	10YR5/6	0.4	0.4	8.5
2Bt	93–150	10YR5/6	0.2	0.2	4.6

oyster shells, along with some animal bones and ash. The soils had very high levels of carbonate, as well as organic carbon contents as high as 7.5%, and levels of citric acid-extractable P in excess of 1000 mg kg^{-1}. Most, but not all, of the soils met the criteria for an anthropic epipedon, as originally defined in U.S. Soil Taxonomy (Soil Survey Staff 1975). The high levels of base saturation meant that the shell-midden soils were classified as Alfisols, in contrast to the native Ultisols from which they were formed. Shell-middens are characteristically higher in exchangeable Ca and pH (pH > 7) than local natural soils, but they can be lower in extractable P (Kaufman and James 1991; Beck 2007; Eberl et al. 2012). Sawbridge and Bell (1972) noted a decrease in total organic carbon and nitrogen with increasing age in Canadian shell-midden soils. This can be attributed to soil degradation as a result of human influence, as is often seen in modern agricultural soils. Shell-midden soils have been likened to Terra Preta soils because they have been shown to have long-term impacts on local vegetation (Cook-Patton et al. 2014; Vanderplank et al. 2014).

"Midden-mounds" are a common type of archaeological site found along streams in the Ouachita Mountains of Arkansas and Oklahoma, USA. The presence of fire-pits and hearths, human burials, remnants of structures, and artifacts dating from the late Archaic to early Woodland archaeological periods (~3000–200 BC), suggests that midden-mounds were sites of human habitation (Holliday 2004). They are characterized by one to two meter-thick, dark gray to black (10YR2/2 to 10YR4/1) topsoils, which contain 1–2% organic C, and which are often described as having a "greasy" feel. Pettry and Bense (1989) studied some 300 year-old sandy anthropogenic soils developed on midden-mounds in northeastern Mississippi, USA, where natural soils were generally classified as Typic Hapludults. Selected properties of a representative midden-mound soil are compared with those of a natural soil in Table 8.2. This pedon had a 113 cm-thick dark reddish-brown (5YR3/2) *A horizon, with a "greasy" feel and well developed granular structure, which contained abundant pottery sherds, chert fragments, stone points and tools, fire-cracked rock and charcoal. The midden-mound soils were generally enriched in organic matter, exchangeable bases, and phosphorous, and had wider C/N ratios and a higher CEC, than the natural soil. They had abundant evidence of biological activity, pH levels of 5.5–6.0, compared with 5.2 or less in natural soils, and they contained 1–5% charcoal by volume. Overall, the midden-mound soils had many distinguishing properties, and they qualified as anthropic epipedons, according to the original definition in U.S. Soil Taxonomy (Soil Survey Staff 1975). The high

Table 8.2 Properties of a natural soil compared with a 300 year-old midden-mound soil (Compiled from Petrry and Bense 1989)

Horizon	Depth (cm)	Color	pH	Organic C (%)	Total N (%)	C/N	Citric acid-ext P (mg kg^{-1})	Ca (cmol kg^{-1})	Mg (cmol kg^{-1})	K (cmol kg^{-1})	Na (cmol kg^{-1})	CEC (cmol kg^{-1})
Midden Mound soil												
*A1	0–10	5YR3/2	5.3	1.9	0.11	16.9	203	11.3	1.3	0.4	0.03	28.6
*A2	10–51	5YR3/2	6.0	1.0	0.05	22.1	218	12.5	0.8	0.1	0.03	24.9
*A3	51–75	5YR3/3	5.7	0.9	0.05	18.9	408	12.9	0.8	0.1	0.04	26.3
*A4	75–87	5YR3/2	5.8	1.1	0.05	21.1	274	15.0	0.7	0.1	0.03	28.0
Natural soil												
Ap	0–12	10YR4/3	4.9	0.6	0.09	7.0	80	0.7	0.7	0.2	0.07	11.4
Bw	12–37	10YR5/3	4.8	0.2	0.04	4.2	–	0.6	0.3	0.1	0.03	7.5
Bg	37–50	10YR5/1	4.7	0.1	0.03	3.4	–	0.7	0.4	0.1	0.04	6.5
C	50–130	10YR6/1	4.7	0.1	0.01	5.0	–	0.8	0.4	0.1	0.05	6.9

Horizons modified from original description

base saturation also meant that they would be classified as Alfisols, in contrast to the natural Ultisol terrain of northern Mississippi.

The definition of the anthropic epipedon has changed several times in U.S. Soil Taxonomy, and the current definition no longer has the requirement of elevated P content (Soil Survey Staff 2014). Midden-soils are probably classified as Anthropic or Anthraltic subgroups of various soil orders. They are probably classified primarily as Hortic Anthrosols, according to the current definitions in the WRB (IUSS Working Group 2015).

8.3 European Dark Earth Soils

The term "Dark Earth" was coined by archaeologists in 1912 for thick, dark-colored, organic-rich, homogeneous, soil-like materials found beneath the oldest parts of London, England. No detailed pedological studies are known and the subsoil is unspecified, but Dark Earth apparently represents the *A horizon of one or more paleosols which overlie structures and strata dating from the late Roman period, and which are overlain by Medieval archaeological strata. Similar Dark Earth materials were subsequently found in other European cities with ancient histories dating back to Roman or Medieval times, such as Moscow (Russia), Paris (France), Brussels (Belgium), and Florence (Italy). Dark Earth initially attracted much archaeological interest in England because its origin was thought to bear on the history of London during the "Dark Ages," i.e. a time during the early Middle Ages for which written historic records are sparse. Unfortunately, the origin of Dark Earth has been difficult to decipher, and interpretations of its origin have proved controversial.

In England, where Dark Earth locally overlies the buried ruins of Roman buildings (Fig. 8.2a, b), and is overlain by Medieval archaeological strata post-dating the Anglo-Saxon period, it is thought to generally range in age from the 5th to late 11th century (Courty et al. 1989; Macphail et al. 2003; Nicosia and Devos 2014). An 8th century coin has been found in Dark Earth, along with pottery dating from between the 3rd century and the 9–10th centuries. In Europe, the Middle Ages or Medieval Period generally lasted from the 5th to 15th century. It began with the fall of the Roman Empire, and merged into the Renaissance Period. The Dark Ages correspond to the early Middle Ages, between the 5th and 10th centuries. In southern England, the Dark Ages began when the Roman legions withdrew ca. AD 410, and ended in 1066 when the Anglo-Saxons were conquered by the Normans. Although it was formed at different times in different places, Dark Earth may collectively span a time interval of ~400–600 years or more in southern England corresponding approximately to the Anglo-Saxon Period.

Two archaeological Dark Earth stratigraphic units were distinguished in London (Courty et al. 1989), a lower "pale Dark Earth" unit, and an upper "Dark Earth" unit (Fig. 8.3). They overlie architectural remnants of Londinium, a Roman settlement founded on the modern site of London in AD 43. The pale Dark Earth unit was

Fig. 8.2 Dark Earth exposed at archaeological site in Leicester, England: a Excavations exposing archaeological strata probably dating from the early Anglo-Saxon period (5th–6th century); b Dark Earth overlying remnants of a rock wall dating from Roman period. Photo by University of Leicester Archaeological Services

found resting on relict Roman floor levels, contained Roman coins and burials, and was cross-cut by later Roman features. It showed little evidence of biological activity or anthroturbation, and was thought to be composed of debris from the burning, collapse and decay of Roman buildings. It also contained midden materials in the form of charcoal, shell, bone, plaster and brick. The overlying Dark Earth unit proper was found overlying mid-Saxon (ca. AD 700) floors locally in London. It was typically 20–90 cm thick, but ranged up to 2 m in thickness, and was characterized by a relatively uniform blackish color ranging from (7.5YR5/1 to 10YR2/1; moist), a moderate pH (6.6–7.5), minor carbonate (<10%), and 1–3% or more organic carbon (Table 8.3). The dark color was due partly to an abundance of charcoal and charred plant remains, which are often macroscopic. Dark Earth impacted by bone, feces or plant decomposition had an elevated P content as high as 1.6–2.6%. Phosphorous-rich Dark Earth contained abundant microscopic vivianite crystals. Artifacts were abundant and were mainly comprised of waste building materials (brickearth and mortar), urban house waste, manure, ceramic tile, and other non-farm debris indicative of human occupation (e.g., oyster shell, bone,

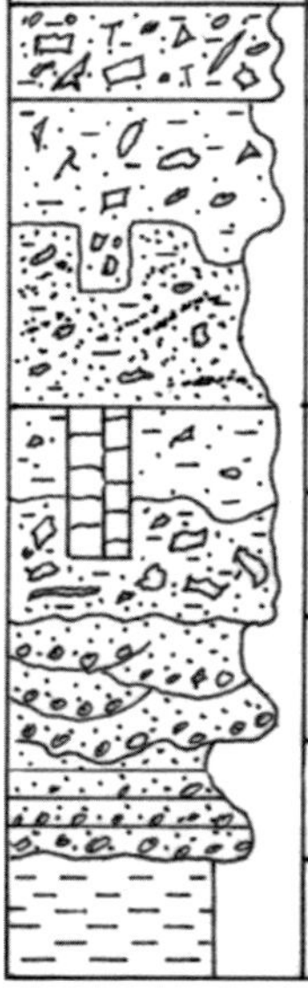

8.3 European Dark Earth Soils

Lithofacies	Description	Age
	Modern building rubble; household artifacts	AD 1700-1900
	Medieval artifacts (sherds) and evidence of post-Anglo Saxon human occupation	AD 1100-1500
	Dark Earth: black, homogeneous soil containing urban household wastes (bone, pottery, shells); heavily bioturbated (earthworms); charcoal	~AD 410-1066
	Pale Dark Earth: contains Roman coins, burials	
	Roman period walls, floors, building debris and artifacts	AD 43-410
	Unnamed sand and gravel deposits comrising terraces of River Thames	Pleistocene
	Bagshot Formation: Deposits of red sand overlying estuarine sand and glauconitic gravel	Eocene
	London Clay: dense silty clay containing brackish water-marine fauna	

Fig. 8.3 Generalized geologic column for London, England showing archaeostratigraphic characteristics of Dark Earth paleosols (Courty et al. 1989; Paul 2016)

pottery). Dark Earth often contained extensive evidence of pedoturbation, particularly in the form of earthworm burrows, and occasionally some phytolithic evidence for agricultural activity. It was characteristically homogeneous, and the dark color made the identification of archaeological features difficult. However, excavations showed that it occasionally contained weak evidence of internal layering or microstratification. Possible tip lines, spreads of concentrated artifacts, and upsection changes in pottery assemblages suggest that Dark Earth formed by vertical accretion.

Excavations have also revealed the presence of an occasional pit-house, i.e. a building that was partly dug into the ground and covered by a roof. Pit-houses, also known as grubenhäuser or grubhuts, were built in many parts of northern Europe between the 5th and 12th centuries A.D. They are believed to have been used as dwellings, to store food, or for other functions. In an archaeological context, an ancient pit-house is usually represented by a dug-out hollow in the ground, along with any postholes that were used to support the roof. The association of pit-houses with sites in London dated by coins and building materials to the 3rd and 4th centuries, has been widely interpreted to indicate that city life in Londonium had become a time of "impoverishment, decay and ruralization of the urban context" (Macphail et al. 2003) toward the end of the Roman period when perhaps many of the wealthier Romans had left the city. The dark, homogenized fabric of Dark Earth was interpreted to reflect mixing of charcoal-bearing domestic wastes with the

Table 8.3 Profile description of Dark Earth paleosol from London, England

Layer	Depth (cm)	Description
*Ab	0–70	Very dark gray (10YR3/1) sandy clay loam; weak coarse subangular blocky structure; 20% coarse fragments and artifacts (charcoal, oyster shell, bone, pottery, brickearth, mortar fragments); common root channels and earthworm burrows; clear smooth boundary
*Cb	70–85	Very dark grayish-brown (10YR3/2) sandy loam; massive structure; firm; abundant brickearth and mortar fragments; few roots channels and earthworm burrows; abrupt smooth boundary
2*Cb	85–95	Strong brown (7.5YR5/6) silt (brickearth); massive structure; abrupt smooth to wavy boundary. Represents Roman earthworks, i.e. floor, wall or mortar foundation of building
3Bgb	95–130+	Light brown (7.5YR6/6) gravelly sand with common Fe concentrations as nodules; massive; few artifacts (charcoal); common root channels

All colors moist. *Note* horizon designations modified from Courty et al. (1989)

remains of decayed pit-huts through bioturbation. Biological activity was possibly enhanced by horticultural activities (Courty et al. 1989).

Early workers thought that Londinium had become an abandoned derelict city after the withdrawal of the Roman legions, and "Dark Earth" was equated with "Dark Ages." The apparent homogeneity of Dark Earth was interpreted as representing some unique soil-forming process linked to urban decline and deterioration. However, it is now thought that habitation of Londinium continued after abandonment by the Romans, perhaps with relatively little decline in population. For example, Dark Earth is sometimes an organic-rich, waterlogged soil containing human excrement probably representing refuse materials that were continuously dumped in a densely populated urban environment. Recent studies have shown that Dark Earth locally contains such domestic habitation features as postholes, fireplaces and interior floors. A rhythmic layering can sometimes be seen suggesting formation by repeated additions of wastes. These features have been interpreted to indicate that Dark Earth was formed partly as an urban midden deposit. Northern London and surrounding areas are underlain by the London Clay (Paul 2016), which is comprised of highly fossiliferous, argillaceous, marginal-marine Eocene strata with high shrink-swell potential. The London Clay is very impermeable, especially when plowed, and very poorly suited for agriculture without improvement. As discussed in Chap. 7, plaggen agriculture was widely practiced in northern Europe during the early Middle Ages (A.D. 500–1000), which coincides with the time of Dark Earth formation in London. Given the evidence for vertical accretion, and the obvious fact that Roman ruins were buried by human-additions of earth materials, it seems possible that Dark Earth may represent a plaggic paleosol created by Anglo-Saxon farmers out of the need to improve the arability of local soils in Londonium. It was also common practice to build homes with thatch roofs composed of sod. These houses lacked a chimney, hence the thatch would become impregnated with soot from smoky fires that burned inside them throughout the winter. These blackened roofs were typically removed and replaced during the spring, and the sooty sods spread on fields as an agricultural amendment. In Abinger, England where Darwin carried out his famous earthworm experiment (Fig. 1.3), Dark Earth is overlain directly by "vegetable mould" (topsoil) formed from earthworm casting activities. This is consistent with the hypothesis that Dark Earth represents an anthropogenic *A horizon formed by plaggen agriculture.

Other paleosols of Dark Earth with characteristics similar to those in London are found in Paris, France (Nicosia and Devos 2014). Paris was founded by the Gauls of the Parisii tribe who settled there between 250 and 200 BC, and founded a fishing village either on an island (Ile de la Cité) in the River Seine, or on its south bank, which evolved into the modern city of Paris. Julius Caesar conquered the territory in 52 BC, and founded the Roman city of Lutetia on the earlier settlement. Paris existed as a regional center during the Roman Period, and prospered during the Middle Ages. Hence, Dark Earth in Paris is roughly the same age as that in London, and thought to date from the 4th to the 11th centuries. The deposits generally range in color from gray (7.5YR5/1) to black (10YR2/1), contain abundant fragments of megascopic charcoal and burnt plant materials, have high C/N ratios, and usually have a high P content (Borderie et al. 2015). Dark Earth in Paris contains approximately 25% carbonate, hence it is much more calcareous than that in London. Artifacts are primarily waste building materials or materials indicative of domestic activities.

Devos et al. (2013) studied Dark Earth paleosols in Brussels, Belgium. Three paleosols were present in a trench dug adjacent to the ancient palace of Hoogstraeten which spanned an interval of ~650 years from A.D. 1010 to 1660. The Dark Earth ranged in color grayish-brown (2.5Y5/2) to gray (2.5Y6/1), and contained an artifact assemblage comprised of brick fragments, bones, charcoal and abundant phytoliths. Data for several of the paleosol layers in one trench are shown in Table 8.4. One of the paleosols had an organic carbon content of 2.2%, but the others were less than 0.3%, although several layers in another trench had organic carbon contents of 0.6–3.7%. Total P levels often exceeded 1000 mg kg^{-1} and the base saturation was ~75%. Several of the paleosols had a plow layer, and there was botanical evidence indicative of agricultural activities such as a grassland (pasture) cover and cereal crops. These was also evidence of manuring, a higher degree of biological activity than in natural soils, and for incremental additions of domestic wastes such as ash, charcoal, charred seeds, bone and ceramics. Overall, the Dark Earth horizons were interpreted as human-made horticultural (garden) soils. Farming pre-dated construction of the palace during the 12th century, and suggests a period of agricultural activity early in the urban history of Brussels which was perhaps similar to that in London.

Dark Earth soils are buried stratigraphic units, therefore they are classified as paleosols and not amenable to the same protocol used to classify surface soils. They resemble Hortic Anthrosols, and if they were found at the surface they might be classified as Plaggic Anthrosols (IUSS Working Group 2015), or as an Anthropic Udorthent, according to U.S. Soil Taxonomy (Soil Survey Staff 2014).

Table 8.4 Chemical characteristics of Dark Earth paleosol from Brussels, Belgium (Devos et al. 2013)

Layer	pH	Organic C (%)	C/N	Total P (mg kg^{-1})	Total N (mg kg^{-1})	Exchangeable bases Ca (cmol kg^{-1})	Mg (cmol kg^{-1})	K (cmol kg^{-1})	Na (cmol kg^{-1})	CEC (cmol kg^{-1})	Base saturation (%)
*4	8.0	2.2	29.1	3889	740	2.99	0.9	0.7	0.3	6.56	75.1
*3	8.2	0.3	12.1	1675	260	0.66	0.4	0.6	0.1	5.33	33.0
*2	7.9	0.2	9.0	980	230	0.45	0.2	0.4	0.1	3.36	34.2
*1	8.1	0.1	7.6	533	150	0.30	0.2	0.3	<0.1	2.64	34.1

8.4 Cemetery Soils

Funerary customs vary widely amongst ancient and modern societies. These variations include differences in the handling of the body, the kind of grave used, the location of burial, and the nature of grave goods buried with the body (Schiffer 1987). The methods of interment which most commonly produce anthropogenic soils are primary burial (inhumation of a corpse in a grave), and secondary or bundle burial. In the case of secondary burial, the dead body was allowed to decay on, in, or above the ground, and then the remains of recently deceased people were later collected and buried together during mound-building or some other type of ceremony.

Evidence for the intentional burials in graves dates back to the late Pleistocene. These include Neanderthal burials from the middle Paleolithic Period about 130,000 year ago in Krapina, Croatia. The oldest known ritualized burial of an anatomically modern human is about 100,000 year old, and was found in a cave at Qafzeh, Israel where human skeletal remains were stained with red ochre, and accompanied by various grave goods. The oldest known grave in Europe is the "Red Lady of Wales" from the late Paleolithic dated at about 29,000 year old. The ancient Sumerians and Chinese were burying their dead in the ground by about 7000 year ago. The ancient Egyptians, Babylonians, Greeks, also buried their dead in earthen graves, and wooden coffins were being used to bury the dead in ancient Scotland and Ireland by about 4000 year ago. In contrast, cremation was the preferred funerary custom used by the civilizations of ancient India and Rome. Many Anglo-Saxon burials and cemeteries are known from England (Longworth and Cherry 1986). In Europe, beginning in about the 7th century AD, the Christian church required all inhumations be done in the consecrated ground of church graveyards. However, this practice began to be discontinued during the 19th century as a result of rapid population growth fueled by the Industrial Revolution. Instead, people began to be buried in grave fields called cemeteries, i.e. tracts of land outside of urban areas specifically designated as a burial ground. There is currently a global lack of space for human burials as many cemeteries and graveyards have been filled to capacity (Hansen et al. 2014).

A grave is typically a rectangular pit that is excavated to a depth of around 180 cm (Fiedler et al. 2012; Majgier et al. 2014). The coffin containing the human remains is placed inside, and the hole is backfilled with the same soil removed from the pit. Even if the soil is piled up on top of the grave, there is inevitably excess soil which is either left on the ground where it was piled up as the grave was dug, or spread about, or carried off to be disposed of elsewhere. Thus, a cemetery or graveyard is expected to be characterized by anthropogenic soils formed in both human-altered and human-transported parent materials.

Table 8.5 compares an undisturbed soil profile with that of a grave anthrosoil from a cemetery in Poland that was used between the 19th century and WWII. The former soil was an Inceptisol characterized by a cambic horizon and a profile thickness of 80 cm. The grave soil was an Entisol 130 cm thick, corresponding to

Table 8.5 Morphological characteristics of a natural soil and a grave soil from a cemetery in Poland, and a grave soil in the USA

Horizon	Depth (cm)	Description
Natural soil (Brunic Arenosol, Poland, ~11,700 year old)		
O	0–1	Decomposed and undecomposed leaves of lilac (*Syringa vulgaris*) and Norway maple (*Acer platanoides*)
A	1–21	Dark grayish-brown (10YR2/2) fine sand con; massive structure; loose; plant roots; abrupt boundary
Bw	21–81	Yellowish-brown (10YR3/4) fine sand; massive structure; loose; abundant roots; iron nodules; diffuse boundary
C	81+	Brown (2.5Y5/3) medium-grained sand containing detrital limestone grains; massive structure; loose; few fine roots
Grave soil ("Necric" Anthrosol, Poland, ~70 year old)		
*O	0–2	Decomposed and undecomposed leaves of Norway maple (*Acer platanoides*) and Scots pine (*Pinus sylvestris*)
*A	2–12	Dark brown (10YR3/3) medium-grained sand; massive structure; loose; abundant roots; abrupt boundary
*C1	12–92	Dark brown (10YR3/3) medium-grained sand; massive structure; loose; abundant roots; diffuse boundary
*C2rhu	92–132	Dark brown (10YR3/3) medium-grained sand; massive structure; loose; residual accumulation of organic matter derived from decomposed human remains; common artifacts (coffin parts, bones, undecomposed textile materials; few roots; abrupt boundary. "Necric" horizon
2C	132+	Brown (2.5Y5/3) medium-grained sand; massive structure; loose; no roots
Grave soil (Anthraltic Udorthent, Connecticut, USA, 112 year old)		
*A	0–40	Very dark brown (10YR2/2) fine sandy loam; pH 5.9.
*Bw	40–65	Yellowish-brown (10YR5/6) fine sandy loam; pH 5.9
*BC	65–96	Dark yellowish-brown (10YR3/4) fine sandy loam; pH 5.6
*Crhu	96–180	Dark brown (10YR3/3) fine sandy loam; pH 5.6. "Necric" horizon
2C	180–200	Olive brown (2.5Y 4/4) fine sandy loam; pH 5.6–5.9

Data compiled from Surabian (2012) and Majgier et al. (2014). *Note* moist colors

the depth of excavation for inhumation. It was characterized by a dark brown (10YR3/3) *A horizon which was not significantly different from the undisturbed soil, overlying a *C1 horizon with the same color corresponding to backfilled soil materials. The *C2 ("necric") horizon was also 10YR 3/3 (moist colors), and contained wood and metal parts remaining from the coffin, along with human remains in the form of bone, bits of clothing and other personal items. The brown color of the "necric" horizon was due in part to the decomposition of wood and corroded metallic coffin parts (Majgier et al. 2014). The data in Table 8.6 show that the "necric" horizon of the grave soil is enriched in organic carbon, and total N and P, compared with the undisturbed cemetery soil. Figure 8.4 shows the profile of a 112 year-old grave soil in Connecticut, USA. The anthropogenic soil profile

8.4 Cemetery Soils

Table 8.6 Properties of a natural soil and a grave soil from a cemetery in Poland (Compiled from Majgier et al. 2014)

Horizon	Depth (cm)	pH	Organic matter (%)	Total N (%)	C/N	Sand (%)	Silt (%)	Clay (%)	Total P (mg kg^{-1})
Natural soil									
A	0–20	7.6	0.5	0.010	30	98.2	1.8	0.0	62
Bw	20–80	7.9	0.4	0.012	20	98.2	1.0	0.8	66
C	80+	8.6	0.2	0.017	6	99.4	0.6	0.0	140
Grave soil									
*A	0–10	7.6	4.7	0.044	63	98.8	1.2	0	186
*C1	10–32	8.0	0.5	0.047	6	98.9	1.1	0	142
*C2rhu	32–52	7.8	4.2	0.066	37	99.5	0.5	0	380
2C	52–100+	8.6	0.1	0.015	5	99.5	0.5	0	328

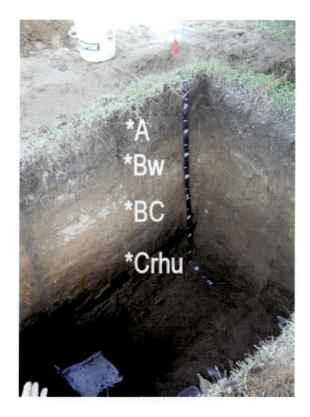

Fig. 8.4 Profile of a 112 year-old grave soil in Connecticut (USA). Note dark colored "necric" horizon in bottom of pit, formed by the residual accumulation of organic matter derived from the decomposition of wooden coffin and associated human remains. Scale in 10 cm intervals. Photo by USDA-NRCS

(Surabian 2012) consisted of a relatively thick *A horizon, a cambic *B horizon, and *BC horizon which was transitional to a dark brown (10YR3/3) *C ("necric") horizon. The "necric" horizon was human-altered parent material enriched in

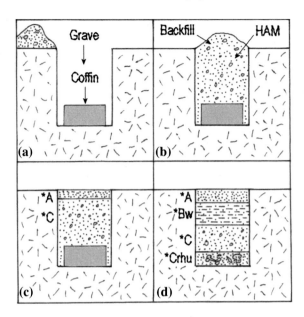

Fig. 8.5 Genesis of a necric horizon: **a** Pit is excavated and coffin containing human remains is placed inside, **b** Hole is backfilled with soil (human-altered parent material is created for new soil), **c** Over time, *A horizon develops via melanization creating an *A-*C profile in human-altered material, **d** After the passage of more time, thicker *A and *Bw horizons form. Artifact-bearing necric (*Crhu) horizon develops as a result of decomposition of wooden coffin, and residual accumulation of organic matter derived from biodecomposition of human corpse

organic matter resulting from the burial, and contained artifacts in the form of bones and coffin parts (wood, nails and handles). Thus, the origin of a profile of a grave soil is unique. In contrast to agricultural anthrosoils, in which an overthickened *A horizon results from human additions of organic matter to the ground surface, a grave soil has an organic-rich *subsurface* horizon which forms *in situ* by the residual accumulation of humus derived from the biodecomposition of intentionally buried human remains and organic artifacts (Fig. 8.5). This unique horizon is referred to informally here as a "necric" horizon, and designated as a *Crhu horizon (rh, residual accumulation of humus; u, artifacts). A "necric" horizon is an organically-enriched subsurface layer of HAM containing burial-related artifacts, and characterized by elevated levels of total phosphorous. It has been referred to previously as a "nekric" horizon (Burghardt 1994; Charzynski et al. 2013). An archaeological excavation of a Civil War grave in northern Virginia, USA was witnessed by the author. The cadaver and wooden coffin were completely decomposed after 115 years of weathering, similar to the time span in Connecticut discussed above. In the absence of a coffin, a "necric" horizon could conceivably form even more rapidly given that cadavers buried in soil can become skeletonized in as little as one or two years (Surabian 2012), and the normal decay time of the human body is generally considered to range from about 25 to 70 years.

No grave soils are known to have been classified formally in the United States according to Soil Taxonomy. Hypothetically, they will be classified as Anthraltic subgroups, probably of Entisol and Inceptisol suborders and great groups. In Germany and Poland they have been referred to as Necrosols (Burghardt 1994; Charzynski et al. 2013; Majgier et al. 2014), and in Russia as Necrozems (Prokofyeva et al. 2011). Grave soils are classified generically as Anthrosols in the WRB (IUSS Working Group 2015), although the recognition of the distinctive "necric" horizon suggests that a new category of Necric Anthrosol could perhaps be added to the WRB in the future.

8.5 Burial Mound Soils

Earthen mounds built by ancient societies are common throughout the northern hemisphere, particularly in the midwestern and southeastern United States, northern Europe, the middle East (e.g., Turkey), and southeastern Asia (China, Japan, Korea). A few sites are known from South America (e.g., Lombardo et al. 2015), Africa and Australia. These mounds, also known as tumuli, barrows, cairns and kurgans, were sometimes apparently used locally only for habitation, or purely for ceremonial purposes, but most are burial mounds. There are an estimated 100,000 burial mounds in southern Scandinavia and northern Europe, where they generally date from the late Neolithic to the early Medieval Period ($\sim$3700 BC to AD 1000). Burial mounds in Europe were often constructed over a single original grave, and then other graves may have been added as new mound layers were deposited (Dreibrodt et al. 2009). Some show signs of agricultural activity during phases of mound construction. In North America, burial mounds are found across 16 states in the U.S., and date from the late Archaic period to the time of European contact ($\sim$3500 BC to AD 1500). North American Indian mounds were often used as sites for secondary burials. They were constructed by hand using baskets filled with soil (Sherwood and Kidder 2011), and are therefore internally layered (Fig. 8.6). Burial mounds generally vary from a few meters to 10 m or more in height, and may be several tens to a few hundred meters in length. They vary from flat-topped pyramids or platform mounds, to rounded cones and elongated ridges. Locally in the U.S., effigy mounds are shaped like animals.

There are some burial mounds that have been abandoned and left relatively undisturbed for 1000 years or more. This has provided soil scientists with an opportunity to study pedogenesis in human-transported material on a very long time scale. Bettis (1988) studied anthropogenic soils on several mound complexes dating from the late Woodland period in northeast Iowa, USA, that were thought to have been used for secondary burials and ceremonial purposes. Iowa has an udic moisture regime with an annual precipitation of $\sim$91 cm. A representative profile of a 1650 year-old burial mound soil is compared with that of a local natural soil (Typic Hapludalf) in Table 8.7. Mound soils had thinner but darker ^A horizons and had cambic ^B horizons, in contrast to the argillic horizons in the natural soils of the area. The mound-top soils had attained organic carbon and total phosphorous

Fig. 8.6 Archaeological stratigraphy of Indian burial mound in eastern Tennessee, USA. Spolic Technosol formed in human-transported material on top of mound can be seen in *upper right*. From Anderson et al. (2013)

Table 8.7 Morphological characteristics of Indian burial mound soils in the USA

Horizon	Depth (cm)	Description
Natural soil (Typic Hapludalf, Iowa)		
A	0–7	Very dark brown (10YR2/2) silt loam; weak medium granular structure; clear smooth boundary
E1	7–21	Dark brown (10YR3/3) silt loam; moderate fine platy structure; clear smooth boundary
E2	21–41	Dark yellowish-brown (10YR4/4) silt loam; moderate medium platy structure; gradual smooth boundary
BE	41–69	Dark yellowish-brown (10YR4/4) silt loam; moderate medium subangular blocky structure; gradual smooth boundary
Bt1	69–99	Dark yellowish-brown (10YR4/4) silt loam; moderate medium platy structure; gradual smooth boundary
Bt2	99–120	Dark yellowish-brown (10YR4/4) silt loam; moderate medium platy structure; gradual smooth boundary
BC	120–150	Dark yellowish-brown (10YR4/4) silt loam; moderate medium platy structure; gradual smooth boundary

(continued)

8.5 Burial Mound Soils

Table 8.7 (continued)

Horizon	Depth (cm)	Description
Burial Mound soil (Typic Dystrochrept, Iowa, ~1650 year old)		
^A	0–6	Very dark gray (10YR3/1) silt loam; weak medium granular structure; abrupt wavy boundary
^E	6–19	Very dark grayish-brown (10YR3/2) silt loam; weak medium platy structure; abrupt wavy boundary
^Bw	19–27	Brown (10YR4/3) silt loam; moderate medium subangular blocky structure; gradual wavy boundary
^C1	27–50	Dark yellowish-brown (10YR4/4) silt loam; massive structure; abrupt smooth boundary
^C2	50–61	Yellowish-brown (10YR5/3) silt loam; massive structure; abrupt smooth boundary
^C3	61–69	Brown (10YR4/3) silt loam; massive structure; abrupt smooth boundary
Burial Mound soil (Typic Hapludult, West Virginia, ~2100 year old)		
^A	0–6	Very dark grayish-brown (10YR3/2) silt loam; moderate medium granular structure; abrupt wavy boundary
^E	6–8	Yellowish-brown (10YR5/3) silt loam; weak fine granular structure; abrupt boundary
^BE	8–17	Yellowish-brown (10YR5/4) silt loam; moderate fine subangular blocky structure; clear smooth boundary
^Bt1	17–46	Dark yellowish-brown (10YR4/4) silt loam; weak fine prismatic and moderate medium subangular blocky structure; clear wavy boundary
^Bt2	46–81	Dark yellowish-brown (10YR4/4) silt loam; moderate medium prismatic and moderate medium subangular blocky; gradual smooth boundary
^Bt3	81–106	Dark yellowish-brown (10YR4/4) silt loam; weak coarse prismatic and weak medium subangular blocky structure; gradual smooth boundary
^BC	106–150	Dark yellowish-brown (10YR4/4) silt loam with common distinct brown (7.5YR4/4) mottles; weak coarse subangular blocky structure; diffuse smooth boundary
^C	150–200	Dark yellowish-brown (10YR4/4) silt loam; massive structure; gradual smooth boundary

Data compiled from Bettis (1988) and Cremeens (1995). *Note* moist colors

contents comparable to those of natural soils. The results showed that a Typic Dystrochrept with an ^A-^E-^Bw profile and weak structure had developed on the mounds in ~1650 years. Cremeens (1995) studied soils developed on a 2100 year-old early Woodland period mortuary mound in West Virginia, USA, a region with an udic moisture regime with an annual precipitation of ~112 cm. A representative profile is given in Table 8.7. Mound-top soils in this area had well developed proto-argillic horizons, and some were characterized by weakly developed argillic horizons and thus classified as Typic Hapludults, similar to local natural soils. These results are consistent with other studies suggesting that cambic

Table 8.8 Time required to form cambic and argillic horizons in humid-temperature climate of the United States based on alluvial and Indian burial mound soils

Location in USA	Type of site	Type of Horizon	Age (year B.P.)	References
Pennsylvania	Alluvium	Cambic	200–2000	Bilzi and Ciolkosz (1977)
Tennessee	Burial Mound	Cambic	1000	Ammons et al. (1992)
Iowa	Burial Mound	Cambic	1650	Bettis (1988)
West Virginia	Burial Mound	Proto-Argillic	2100	Cremeens (1995)
Iowa	Burial Mound	Proto-Argillic	2500	Parsons et al. (1962)
Louisiana	Burial Mound	Argillic	4300	Saunders and Allen (1994)

horizons are well expressed in anthroprogenic soils ∼500–2000 years old, proto-argillic horizons are evident after ∼2000–2500 years, and argillic horizons are formed in approximately 2500–3000 years (Table 8.8).

Burial mound soils are classified as Spolic Technosols, according to the WRB. As discussed above, burial mound soils in the United States have been classified as Entisols, Inceptisols and Ultisols. Mound soils affected by human habitation, and thus characterized by an anthropic epipedon, could perhaps be classified as an Anthropic subgroup of Entisol, Inceptisol or Ultisol, according to the current system of U.S. Soil Taxonomy (Soil Survey Staff 2014), or as a Hortic Anthrosol using the WRB (IUSS Working Group 2015).

References

Ammons JT, Newton JE, Foss DL, Lynn WR (1992) Soil genesis of two Indian mound soils in west Tennessee. Soil Surv Horizons 33:39–45
Anderson DG, Cornelison JE, Sherwood SC (2013) Archaeological excavations at Shiloh Indian Mounds. National Park Service, Tallahassee, FL, 774 p
Beck ME (2007) Midden formation and intrasite chemical patterning in Kalinga, Philippines. Geoarchaeology 22:453–475
Burghardt W (1994) Soils in urban and industrial environments. Z Pflanz Bodenk 157:205–214
Bettis EA III (1988) Pedogenesis in Late Prehistoric Indian mounds, upper Mississippi valley. Phys Geog 9:263–279
Bilzi AF, Ciolkosz EJ (1977) Time as a factor in the genesis of four soils developed in recent alluvium in Pennsylvania. Soil Sci Soc Am J 41:122–127
Borderie Q, Devos Y, Nicosia C, Cammas C, Macphail RI (2015) Dark Earth in the geoarchaeological approach to urban contexts. In: Carcaud N, Arnaud-Fassetta G (eds) French geoarchaeology in the 21st century. Alpha CNRS, Paris, France, p 620
Charnzynski P, Bednarek R, Greinert A, Hulisz P, Uzarowicz L (2013) Classification of technogenic soils according to World Reference Base in the light of Polish experiences. Soil Sci Annual 64:145–150
Cook-Patton SC, Weller D, Rick TC, Parker JD (2014) Ancient experiments: forest diversity and soil nutrients enhanced by Native American middens. Landscape Ecol 29:979–987
Courty MA, Goldberg P, Macphail R (1989) Soils and micromorphology in archaeology. Cambridge University Press, Cambridge, 344 p

References

Cremeens DL (1995) Pedogenesis of Cotiga Mound, a 2100-year-old Woodland mound in southwest West Virginia. Soil Sci Soc Am J 59:1377–1388

Devos Y, Nicosia C, Vrydaghs L, Modrie S (2013) Studying urban stratigraphy: Dark Earth and a microstratified sequence on the site of the Court of Hoogstraeten (Brussels, Belgium). Integrating archaeopedology and phytolith analysis. Quat Intern 315:147–166

Dreibrodt S, Nelle O, Lutjens I, Mitusov A, Clausen I, Bork HR (2009) Investigations on burial soils and colluvial layers around Bronze Age burial mounds in Bornhoved (northern Germany): an approach to test the hypothesis of "landscape openness" by the incidence of colluviums. The Holocene 19:487–497

Eberl M, Alvarez M, Terry RE (2012) Chemical signatures of middens at a Late Classic Maya residential complex, Guatemala. Geoarchaeology 27:426–440

Erlandson JM (2013) Shell middens and other anthropogenic soils as global stratigraphic signatures of the Anthropocene. Anthropocene 4:24–32

Fiedler S, Pusch CM, Holley S, Wahl J, Ingwersen J, Graw M (2012) Graveyards—special landfills. Sci Total Environ 419:90–97

Hansen JD, Pringle JK, Goodwin J (2014) GPR and bulk ground resistivity surveys in graveyards: locating unmarked burials in contrasting soil types. Forensic Sci Intern 237:e14–e29

Holliday VT (2004) Soils in archaeological research. Oxford University Press, Oxford, 465 p

IUSS Working Group WRB (2015) World Reference Base for Soil Resources 2014, update 2015. International soil classification system for naming soils and creating legends for soil maps. World Soil Resources Reports No. 106. FAO, Rome

Kaufman IR, James BR (1991) Anthropic epipedons in oyster shell middens of Maryland. Soil Sci Soc Am J 55:1191–1193

Limbrey S (1975) Soil science and archaeology. Academic Press, London, 384 p

Longworth I, Cherry J (1986) Archaeology in Britain since 1945. British Museum Publication, London, 248 p

Lombardo U, Denier S, Veit H (2015) Soil properties and pre-Columbian settlement patterns in the Monumental Mounds region of the Llano de Moxos, Bolivian Amazon. Soil 1:65–81

Macphail RI, Galinie H, Verhawghe F (2003) A future for dark earth? Antiquity 77:349–358

Majgier L, Rahmonov O, Bednarek R (2014) Features of abandoned cemetery soils on sandy substrates in northern Poland. Eurasian Soil Sci 47:621–629

Nicosia C, Devos Y (2014) Urban Dark Earth. In: Smith C (ed) Encyclopedia of global archaeology. Springer, New York, pp 7532–7540

Parsons RB, Scholtes WH, Riechen FF (1962) Soils of Indian mounds in northeastern Iowa as benchmarks for studies of soil genesis. Soil Sci Soc Am Proc 26:491–496

Paul JD (2016) High-resolution geological maps of central London, UK: comparisons with the London underground. Geosci Front 7:273–286

Pettry DE, Bense JA (1989) Anthropic epipedons in the Tombigee Valley of Mississippi. Soil Sci Soc Am J 53:505–511

Prokofyeva TA, Martynenko IA, Ivannikov FA (2011) Classification of Moscow soils and parent materials and its possible inclusion in the classification system of Russian soils. Eurasian Soil Sci 44:561–571

Saunders JW, Allen T (1994) Hedgepeth Mounds, an Archaic mound complex in north-central Louisiana. Am Antiq 59:471–489

Sawbridge DF, Bell MAM (1972) Vegetation and soils of shell middens on the coast of British Columbia. Ecology 53:840–849

Sherwood SC, Kidder TR (2011) The DaVincis of dirt: geoarchaeological perspectives on Native American mound building in the Mississippi River basin. J Anthrop Archaeol 30:69–87

Schiffer MB (1987) Formation processes of the archaeological record. University of New Mexico Press, Albuquerque, NM, 428 p

Smith G (1986) The Guy Smith interviews: rationale for concepts in soil taxonomy. In: Forbes TR (ed) USDA soil management Service Technology Monog. 11, Cornell University, Ithaca, NY

Soil Survey Staff (1975) Soil taxonomy: a basic system of soil classification for making and interpreting soil surveys. USDA-SCS Agric. Handbook 436, U.S. Gov. Printing Office, Washington, DC

Soil Survey Staff (2014) Keys to soil taxonomy, 12th edn. U.S. Department of Agriculture, Natural Resources Conservation Service, 372 p

Surabian D (2012) Preservation of buried human remains in soil. USDA-NRCS, Tolland, Connecticut, 54 p

Vanderplank SE, Mata S, Ezcurra E (2014) Biodiversity and archaeological conservation connected: aragonite shell middens increase plant diversity. Bioscience 64:202–209

Chapter 9
Mine-Related Anthropogenic Soils

Mine-related anthropogenic soils are primarily associated with modern landscapes created by the surface mining of coal. They are classified as Spolic Technosols having been formed in HTM containing technogenic artifacts in the form of mine spoil. Mine spoils are comprised of sedimentary rock fragments that have been fractured and rapidly exposed by blasting and excavation. Spoils are composed predominantly of coarse fragments, and are characterized by unusually large voids not seen in natural soils. Some mine spoils contain significant amounts (30–60%) of fine earth fraction depending on blasting intensity. Mine soils generally have ^A–^C profiles. Weakly developed ^A horizons have formed in 1–5 years, and ^AC horizons have developed after 8–13 years of pedogenesis. Cambic-like ^B horizons have formed in 20–50 years, but often do not meet thickness and color requirements for a cambic horizon. Physical weathering of mine spoils is often noticeable within 6 months to a year due to microfracturing caused by the blasting process. Accelerated chemical weathering has been attributed to rapid leaching through large macropores, and to the large surface area of abraded mineral surfaces which are in disequilibrium with surface environmental conditions. Young mine soils can be higher in pH, exchangeable bases, water-soluble Fe, and total P than local natural soils, but weathering of pyrite-bearing rocks may produce sulfuric acid (acid mine drainage). Young mine soils are commonly very high in soluble salts, derived by reaction of acid mine drainage and calcareous rocks. Accumulation of salt may occur within 2–3 years or less, and leaching can occur in <30 years. CEC is initially very low, but increases over time as soil organic matter accumulates. Natural revegetation of mine soils can be retarded by low levels of N, diminished microbial activity, or by a very low pH and the presence of toxic levels of metals generated by acid mine drainage.

9.1 Introduction

Mine-related anthropogenic soils are primarily associated with the surface mining of coal, a carbonaceous rock formed from plant materials which have accumulated under anaerobic conditions in a swamp or bog. Coal deposits are found on every continent, but the largest reserves are in the United States (27%), Russia (17%), China (13%), India (10%), and Australia (9%). There is archaeological evidence in China for surface mining of coal and household usage by $\sim$3490 BC. Flint axes embedded in coal show that it was also being mined in Britain during the Stone and Bronze Ages. Romans were mining British coal by $\sim$AD 200, and used it to heat their baths. Most major British coal fields were being worked more extensively by the 12th century. Coal was being mined in New Brunswick, Canada by the 1600s, and in the U.S. by the 1740s. Coal mining began in many countries during the 1700s, but increased significantly during the 19th and 20th centuries in response to the Industrial Revolution. Coal mining peaked in many places during the early 1900s, and has declined greatly since the 1970s. Many coal mines in Europe closed during late 1900s and early 2000s. Today, the largest coal producers are China, the U.S., India, Australia, and Indonesia.

There are two basic coal mining methods, underground and surface mining (also known as open cast mining). Underground mining was the predominant method used in the past, and it is still the main method used in certain locations, such as China, where more than 90% of coal is extracted by underground mining (Chang-sheng et al. 2009). Surface mining overtook underground mining in the U.S. around 1972, and although Australia produces about 3/4 of its coal by surface mining, the greatest number of surface mines is located in the Appalachian basin of the eastern United States. Other significant areas of surface mining in the U.S. are the Interior region (Illinois, Iowa and Michigan), and the Western region (North Dakota, New Mexico, Colorado, Utah, Wyoming, Montana). There are currently about 538 surface mines in the Appalachian basin, mostly in Pennsylvania, compared with 64 and 52 in the Interior and Western regions, respectively. By comparison, China has a total of only about 21 major surface mines. Most surface mines are located in the eastern U.S. because strip mining started there around 1940. Coal seams are also thinner in the eastern U.S., hence more land surface must be disturbed to recover same tonnage as in the western U.S. It is estimated that as much as 8 million ha have been disturbed by surface mining in the U.S., compared with about 3 million ha in China (Chang-sheng et al. 2009).

There are three main surface mining techniques used for coal extraction (Toy and Hadley 1987): (1) contour or strip mining, (2) mountaintop removal, and (3) areal mining. Open pit mining, which is the most common method used for the extraction of non-coal mineral resources, is rarely used for coal mining. **Strip mining** is typically used where coal is found in more or less horizontal layers in hilly or mountainous terrain. It is generally carried out in four steps (Fig. 9.1). First, the site is prepared by removal of vegetation and soil cover, if present. Second, the overburden rock above the coal seam is drilled and blasted while following the contour of

9.1 Introduction

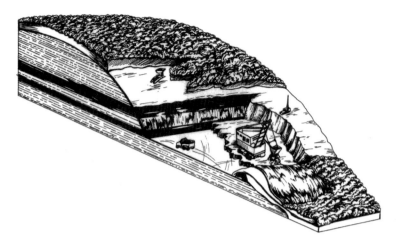

Fig. 9.1 Typical strip mining operation in Appalachian Mountains. Strip mining is carried out in three steps: *1* site is prepared by removing vegetation and soil, *2* overburden is removed by drilling and blasting, *3* overburden is removed above coal seam, and *4* coal seams is excavated and carried away by truck and train (Grim and Hill 1974)

the land surface, often for many kilometers. Third, the overburden is removed and a terrace-like bench is created with only the coal seam remaining. Fourth, the coal is excavated, loaded into trucks, and carried to sites downslope where further processing may take place, and the coal is ultimately loaded into train cars.

Mountain-top removal is usually carried out in association with strip mining of flat-lying, homoclinal strata, and involves removal of entire mountain tops by drilling and blasting along a series of parallel cuts. The overburden is removed down to the coal, which is then extracted as in strip mining. Mountain-top removal produces a level plateau or a gently rolling upland surface. During strip mining and mountain-top removal, excess spoil (i.e. non-coal waste rock fragments) may be used to fill in natural depressions, pushed onto the side slopes or into the heads of streams valleys, or simply piled onto the tops of ridges and spurs.

Areal mining is used in flat to gently undulating terrain, and is the most commonly used method in the Interior and Western regions of the U.S. After removal of the vegetation and soil, an initial cut is made down to the coal, and the spoil is dumped onto a stockpile to one side of the planned mine. This cut is referred to as a box cut and is extended across the property to the limits of the planned mine, with the coal being removed as in a strip mining operation. Additional cuts are then made parallel to the first, with the spoil from each succeeding cut being placed in the previously mine-out pit. Each cut may be several hundreds or thousands of meters in length, and the final cut may be more than a kilometer from the initial cut. The resulting land mined by areal stripping resembles an enormous washboard.

The "shoot-and-shove" method was the most common type of strip mining technique used in the Appalachians before the 1970s (Fig. 9.2). Blasting and

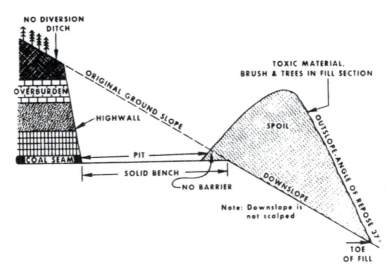

Fig. 9.2 Bench-and-outslope topography created by strip mining prior to implementation of the Surface Mining Control and Reclamation Act (1977) in the United States (Grim and Hill 1974)

excavation were employed to remove rock overburden comprising homoclinal sequences lying above bedded coal seams outcropping along the side of a ridge (Fig. 9.3a). This produced a landscape composed of a high wall, a bench remaining after the coal was removed, and a steep outslope composed of mine spoil which had been bulldozed over the edge of the bench (Fig. 9.3b). Multiple benches were present on many ridges, and benches were often so close together that outslope spoil from a bench above overlapped with that from one below. The Surface Mining Control and Reclamation Act of 1977 introduced new requirements that surface mined lands be restored to approximate original contour, which usually involved covering the exposed high wall with spoil. It also mandated that acid-producing pyritic materials

Fig. 9.3 Strip mining as it was carried out in southwestern Virginia during 1976: **a** "Shoot-and-shove" method produced multiple benches and large-scale devastation of the landscape. **b** Overburden and coal have been removed leaving a narrow bench and "high wall" ~15–20 m in height. Photos by J. Howard

be buried below the surface, and that a healthy, self-sustaining vegetal cover be established and maintained for five years beyond the initial planting, before bond monies posted by the mine operator could be recovered. Mine spoils with suitable properties were allowed to be used as topsoil substitutes because natural soils in some parts of Appalachia were too thin to be used for land reclamation.

9.2 Geologic Setting

There were three main episodes of coal formation during Earth's history: (1) Carboniferous-Permian, (2) Jurassic-Cretaceous, and (3) Paleocene-Eocene. Coal-bearing strata of Carboniferous-Permian and Paleocene-Eocene age are found on all of the continents, whereas coals of Jurassic-Cretaceous age are found in Canada, the U.S., China, and the Commonwealth of Independent States (Thomas 2012). Coals are found in diverse structural settings (continental rift, pull-apart basin, foreland basin, etc.), but there are many lithostratigraphic similarities in coal-bearing stratigraphic sequences worldwide despite being separated by great differences in geologic age and geographic location. This is attributable to the fact that economic deposits of coal were generally formed in similar paleoenvironments. In addition to the presence of carbonaceous materials formed by the carbonization of humic substances, coal-bearing strata are generally characterized by extensive iron-bearing oxide and sulfide mineralization.

Coals are generally found in rhythmic successions of interbedded terrestrial clastic and marine carbonate rocks. The cyclicity of coal-bearing strata was thought to be the result of a transgressive-regressive sequence called a cyclothem. The classic cyclothem graded upward from coal, to fossiliferous, marine, calcareous mudrock or limestone (known as roof rock), and then to non-marine sandstone and siltstone. These strata were overlain by argillaceous or carbonaceous mudrock. Although these stratigraphic relationships are commonly observed in the geologic record, the cyclothem concept has been replaced by facies and sequence-stratigraphic models of coal sedimentation (Holz et al. 2002). It is now well established that the cyclic nature of coal-bearing strata is partly the result of inherently rhythmic coastal sedimentary processes, and partly due to cyclic variations in global climate and sea level. Most economic deposits of coal were formed in deltaic or barrier island depositional settings (Diessel 1992; Thomas 2012). In a typical prograding deltaic system, lithofacies grade sequentially upward from prodelta, to delta-front, lower delta plain, and upper delta plain depositional settings (Fig. 9.4). Coal was formed from peat which accumulated in forested coastal swamps, usually located along rivers, with slow-moving or stagnant fresh, brackish or marine water. Most coal was formed in a lower delta plain setting near mean high tide level where swamps were located around tidal inlets, along the shores of interdistributary bays, and on the levees of distributary channels. Rapid deltaic facies changes can result in multiple or split coal seams, which are often mined simultaneously.

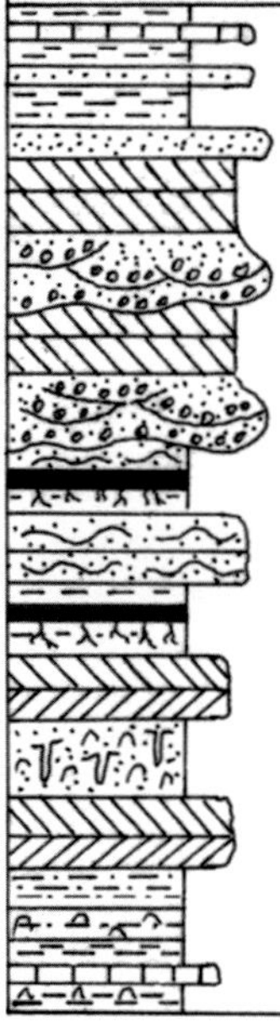

Lithofacies	Description	Interpretation
	Upward-fining arenite with unidirectional cross-beds (point bar); siltstone (levee); mudrock with limestone, coal (overbank)	Upper Delta-Plain Fluvial Complex
	Thick-bedded arenite woth large-scale trough and unidirectional planar cross-beds (interdistributary channel); plant fossils and brackish marine fauna possible	Lower Delta-Plain Interdistributary Channel
	Bioturbated flaster-bedded arenite with bi-directional cross-beds (tidal channel); mudrock with plant fossils, mudcracks, brackish water fauna (tidal flat); coal and underclay with root clasts, plant fossils	Lower Delta-Plain Interdistributary Bay
	Arenite with large-scale, bi-directional planar cross-bedding; bioturbated arenite, vertical burrows common	Lower Delta-Plain, Beach-Bar Complex
	Argillaceous shale and siltstone with brackish water fauna; Lime mudstone with open marine fauna	Pro-Delta Shallow Marine Shelf

Fig. 9.4 Lithofacies assemblages comprising idealized deltaic sedimentary sequence showing typical locations of coal-bearing strata (solid black)

Coal-bearing clastic strata are characterized by considerable Fe-oxide mineralization, primarily goethite, in the form of pore-filling cement, beds, nodules, rip-up clasts, and replacement fossils (Fig. 9.5a). These represent bog iron deposits, which formed when eroded soil particles carrying Fe-oxide coatings were transported by streams and deposited in a swamp. The Fe was solubilized under anaerobic conditions, percolated through the underlying sediments, and was then oxidized and precipitated out when it encountered oxidizing ground or surface waters. Coals and associated carbonaceous shales are characterized by variable amounts of pyrite and possibly other sulfide minerals. This pyrite is partly syngenetic and may be found intimately associated with coal macerals as lamina (Fig. 9.5b), framboids, euhedra, or irregular masses. Syngenetic iron sulfides form as a by-product of sulfate reduction, i.e. the use of SO_4^{-2} by microbes during respiration. Seawater was probably the principal source of the sulfate, hence pyrite is most common in coals which are overlain by roof rocks formed under brackish or marine conditions. Pyrite is also commonly epigenetic, and forms veins and fillings in coal cleats and other fractures.

When surface mining exposes pyrite (and other iron disulfide minerals) to the surface environment, it oxidizes and hydrolyzes to form acid mine drainage according to the following sequence of reactions (Akeil and Koldas 2006):

9.2 Geologic Setting

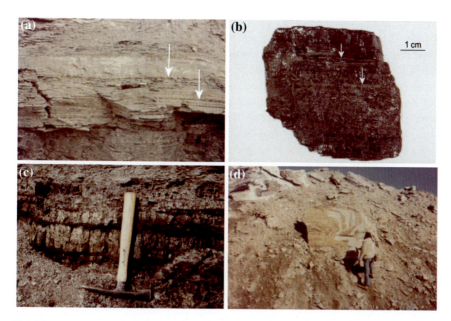

Fig. 9.5 Geologic features of coal-bearing rocks. **a** Goethite interbeds in calcareous siltstone. *Note* salt efflorescence. **b** Pyrite laminae in coal. **c** Dried sulfuric acid and sulfate on coal outcrop. **d** Mine spoil containing significant soil-sized (<2 mm) material is used as a topsoil substitute in steeply sloping terrains where there is little or no natural soil covering. Photos by J. Howard

$$2FeS_2 + 2H_2O + 7O_2 \rightarrow 2Fe^{2+} + 4SO_4 + 4H^+ \quad (9.1)$$

$$4Fe^{2+} + O_2 + 4H^+ \rightarrow 4Fe^{3+} + 2H_2O \quad (9.2)$$

$$4Fe^{3+} + 12H_2O \rightarrow 4Fe(OH)_3 + 12H^+ \quad (9.3)$$

$$FeS_2 + 14Fe^{3+} + 8H_2O \rightarrow 15Fe^{2+} + 2SO_4^{2-} + 16H^+ \quad (9.4)$$

Initial oxidation and hydrolysis of pyrite in reaction 9.1 occurs very rapidly, within weeks or even days of exposure. Reaction 9.2 is the rate-determining step and may require several years to occur under abiotic conditions. However, the conversion of ferrous to ferric Fe is greatly accelerated by various species of bacteria such as *Thiobacillus ferroxidans*. The activity of these extremophile bacteria is actually favored by a pH in the range of 2–3, with minimal activity occurring above a pH of about 6.0. Reaction 9.3 involves hydrolysis of ferric Fe to form solid ferrihydrite, commonly known as "yellow boy" because of the distinctive yellowish-orange color it imparts to the highly acid effluent. Reaction 9.3 is pH-dependent with ferrihydrite only forming at a pH greater than about 3.5, and ferric Fe remaining in solution below pH 3.5. Various sulfate minerals may also form by precipitation such as jarosite ($KFe_3(OH)_6(SO_4)_2$) and schwertmannite ($Fe_8O_8(OH)_6(SO_4)–nH_2O$). In the case of reaction 9.4, additional pyrite is oxidized by excess ferric Fe produced by reaction 9.2. Oxygen is not required for reaction 9.4, which takes place very rapidly, and occurs until the supplies of ferric Fe or pyrite are used up. Acid mine drainage is characterized by very low pH (1.4–2.9), and high sulfate and dissolved Fe contents. It is produced anywhere rocks containing sulfides are disturbed, such as mining of coal, Cu, Ag, Au, Zn, Pb, and U. The same phenomenon is associated with acid sulfate soils, i.e. where pyrite-bearing coastal or estuarine peats are artificially drained usually for agricultural purposes (Kittrick et al. 1982).

Sulfuric acid mine drainage can react with any calcite present as pore-filling cement, or as a detrital framework component, to form gypsum:

$$H_2SO_4 + CaCO_3 + H_2O \rightarrow CaSO_4 - 2H_2O + CO_2 \qquad (9.5)$$

Coal seams coated in dried sulfuric acid and sulfate (Fig. 9.5c), and clastic rocks with an efflorescence of gypsum or other soluble salts, are good field indicators of acid mine drainage.

9.3 Coal Mine-Related Anthrosoils

Mine-related anthropogenic soils are distinctive in that they are derived from unique parent materials in the form of mine spoils (tailings). Mine spoils are anthropogenic sediments comprised mainly of fragments of unweathered sedimentary rock which have been abraded, fractured, and rapidly exposed by blasting and excavation. These blasted rock fragments may be present in a wide variety of particle sizes, ranging from boulders 1–2 m in diameter, to fine silt and clay. Mine spoils are typically composed of $\geq 60\%$ coarse fragments and, if uncompacted, they are characterized by unusually large voids not seen in natural soils. Such voids are creating by loose packing during the backfilling process, and range in size from a few centimeters to 0.5 m in diameter. The voids eventually disappear as a result of settling and compaction of spoil under its own weight. Plant roots have been observed to follow these voids to depths of 50 cm or more. Hence, the presence of large packing voids promotes natural and artificial revegetation of mine spoils. The presence of large voids also may promote drainage and aeration, and thereby accelerate soil formation.

Freshly blasted mine spoil may contain 30–60% soil-sized (<2 mm) material (Fig. 9.5d), depending on drill-hole spacing. The particle size distribution of the fine earth fraction is controlled by rock type such that arenites and siltstones produce spoils with sand to sandy loam, and silt to silt loam, textures, respectively, whereas a mixture of sandstone and siltstone tends to generate spoils with a loamy texture. Overall, the texture of mine spoils is highly variable as a result of the abrupt vertical and horizontal lithofacies changes characteristic of coal-bearing strata. Clay

9.3 Coal Mine-Related Anthrosoils

content is typically low (<15% by weight). The mineralogical composition of mine spoils is also highly variable, both as a result of both lithologic and geochemical facies changes. Arenites and siltstones are generally quartzose, with lesser amounts of feldspar and rock fragments. These rocks may contain minor amounts (<2%) of clay-sized material, with a layer silicate mineral assemblage analogous to that found in soils. However, these clay minerals (e.g., kaolinite and montmorillonite) are geogenic and were formed as a result of diagenesis. The clay fraction of mine spoils may also be compositionally similar to the "rock flour" found in glacial settings, in that mine spoils may contain minerals not normally associated with the clay fraction of soils, such as quartz, feldspar and calcite, which have been produced artificially by blasting. Arenites may have significant primary porosity, and vary in hardness and chemical resistance depending on whether calcite or goethite cements predominate. In the absence of these cements, the rocks may be very hard and resistant as a result of silica cementation. Siltstones are similar in composition to arenites, but often with a greater abundance of mica. Claystones range from argillaceous to highly carbonaceous, and commonly represent the paleosols in which the peat beds formed that were subsequently transformed into coal. Conscientious mine operators may try to bury carbonaceous rocks beneath 1–2 m of spoil, in order to avoid acid mine drainage, but inevitably carbonaceous rock materials are scattered throughout the spoils.

Physical weathering of mine spoils occurs very fast at, or near, the ground surface. Physical disintegration of clasts, especially non-persistent types of mudrock, is noticeable within 6 months to a year. Accelerated physical weathering is attributed to microfractures caused by the blasting process, which are visible in thin sections of mine spoil rock fragments. This is especially true locally where the drill-hole spacing used for blasting may be on the order of 2 m. Many studies have suggested that accelerated physical disintegration results in a measureable decrease in coarse fragments in the upper 15–25 cm, compared with deeper parts of a mine soil profile. It also appears that after a few years there is an "annealing" of large voids, and an evolution into massive structure to a depth of perhaps 100 cm, as a result of settling and infilling by finer material generated by physical weathering (Fig. 9.6).

Subtle differences in mine soil morphology are difficult to detect because of the overall heterogeneity of the spoil materials. Mine soils generally appear to have ^A–^C, ^A–^AC–^C or ^A–^Bw–^C profiles (Table 9.1). Weakly developed ^A horizons about 5 cm thick showing some organic matter accumulation, but with massive or weak granular structure, have been reported in mine soils only 1–2 years old (Daniels and Amos 1985; Haering et al. 1993). ^A horizons may be 5–10 cm thick and have well developed granular structure within about 5 years, and ^AC horizons are reported to have developed within about 8–13 years (Ciolkosz et al. 1985; Haering et al. 1993). Figure 9.7 illustrates the characteristics of an eight year-old mine soil from southwestern Virginia, USA. The soil had a well developed ^A horizon ~5–10 cm thick, characterized by weak to moderate granular and subangular blocky structure. The ^AC horizon was darker than the ^C and had some evidence of granular structure, compared with the underlying ^C1 horizon.

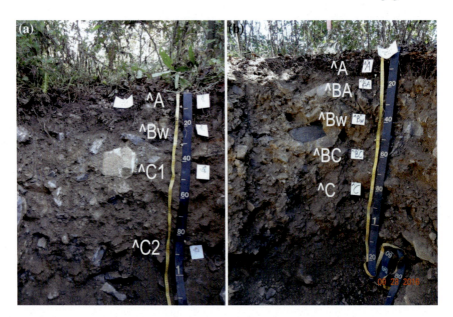

Fig. 9.6 Mine soil profile development in southwestern Virginia, USA after 32 years of pedogenesis: **a** No amendments added; **b** biosolids added as amendment for land reclamation. Megavoids have disappeared in both profiles as a result of compaction and infilling by fine earth fraction. Scale in cm. Photos by W. Lee Daniels

The ^C1 was more compacted, had smaller voids, and contained smaller coarse fragments, than the underlying ^C2 horizon. Cambic and cambic-like ^B horizons are also widely reported from mine soils. These horizons are invariably recognized on the basis of structure, often do not meet thickness and color requirements for a cambic horizon, and tend to be vertically and/or laterally discontinuous. Shafer et al. (1980) reported cambic-like ^B horizons in 50 year old mine soils from Montana that did not meet the thickness requirement. Ciolkosz et al. (1985) also reported cambic-like ^B horizons in 20 year old mine soils from Pennsylvania, and soils with cambic horizons possibly classified as weakly developed Inceptisols in 29 year-old mine soils. Other cambic-like ^B horizons have been described in mine soils about 20 years old from Virginia and West Virginia, and some have met the requirements for a cambic horizon (Haering et al. 2004).

Mine soils can be highly compacted by earthmoving equipment during the backfilling process. Daniels and Amos (1985) compared five year-old mine soils at two adjacent sites in Virginia, USA and found that well vegetated mine soil lacked a densic horizon, whereas mine soil barren of vegetation was highly compacted (>1.8 g cm^{-3}) below 23 cm depth. Densic conditions were still evident after 25 years in some West Virginia mine soils (Table 9.2). Excess compaction can restrict internal drainage resulting in the presence of poorly drained mine soils with gleyed horizons (Table 9.1). Silty mine spoils are also prone to piping and surface

9.3 Coal Mine-Related Anthrosoils

Table 9.1 Morphological characteristics of mine soils from southwestern Virginia. Data compiled from Haering et al. (2004, 2005)

Horizon	Depth (cm)	Description
Well drained soil (Anthroportic Udorthent)		
^A	0–11	Dark grayish-brown (10YR4/2) gravelly loam with thick, vesicular crust; weak fine subangular blocky structure; firable; common fine roots; 20% coarse fragments (sandstone and siltstone); 5% carboliths; strongly acid (pH 5.5); clear smooth boundary
2^Cd	11–46	Dark gray (10YR4/1) gravelly silt loam; massive; very firm; few fine roots along coarse fragment faces; 12% coarse fragments (sandstone and siltstone); 6% carboliths; extremely acid (pH 4.4); clear wavy boundary
3^C	46–100+	Dark gray and yellowish-brown (10YR4/1 and 10YR5/6) very cobbly sandy loam; structureless; firm; no roots; 34% coarse fragments (sandstone); 1% carboliths; very strongly acid (pH 4.7)
Well drained soil (Anthroportic Udorthent)		
^A	0–6	Dark grayish-brown (10YR4/2) very cobbly silt loam; weak very fine to fine subangular blocky structure; very friable; few coarse and common fine roots; 57% rock fragments (siltstone and shale); <1% carboliths; medium acid (pH 5.8); abrupt smooth boundary
^Bw	6–19	Yellowish-brown (10YR5/4) very cobbly loam; weak to moderate medium and fine subangular blocky structure; areas of both friable and firm consistence; common fine roots; 43% rock fragments (siltstone); <1% carboliths; very strongly acid (pH 5.0); clear smooth boundary
^C	19–47	Yellowish-brown (10YR5/6) to brownish-yellow (10YR6/6) cobbly loam, with common coarse gray (10YR5/1) lithochromatic color variegations; massive; firm; few fine roots; 48% coarse fragments (sandstone and siltstone); <1% carboliths; strongly acid (pH 5.3); abrupt smooth boundary
2R	47–70+	Gray (10YR5/1) sandstone bedrock
Very poorly drained soil (Anthroportic Epiaquent)		
^Ag	0–20	Dark gray (10YR4/1) very gravelly sandy loam with 8% reddish-brown (10YR4/4) and 6% yellowish-brown (10YR5/8) Fe concentrations as pore linings, and common light olive-brown (2.5Y5/4) lithochromatic variegations; moderate medium subangular blocky structure; friable; 40% coarse fragments (sandstone, siltstone, carboliths); many medium, fine, and very fine roots; slightly acid (pH 6.2); gradual wavy boundary
^ACg	20–55	Gray (10YR4/2) very gravelly loam, with 6% strong brown (7.5YR5/8) Fe concentrations and 4% dark gray (2.5Y4/1) Fe depletions as pore linings, and common brownish-yellow (10YR6/6) lithochromatic color variegations; weak coarse subangular blocky structure; firm with some friable areas; 60% coarse fragments (sandstone, siltstone, shale, carboliths); many medium, fine and very fine roots; neutral (pH 7.3); gradual wavy boundary
^Cd	55–140+	Gray (10YR4/2) very gravelly loam, with 5% strong brown (7.5YR5/6 and 5/8) Fe concentrations and 3% dark gray (2.5Y4/1)depletions as pore linings, and common brownish-yellow (10YR6/8) lithochromatic color variegations; structureless massive; very firm; 60% coarse fragments (sandstone, siltstone, shale, carboliths); few fine and very fine roots in upper part of horizon; neutral (pH 7.3)

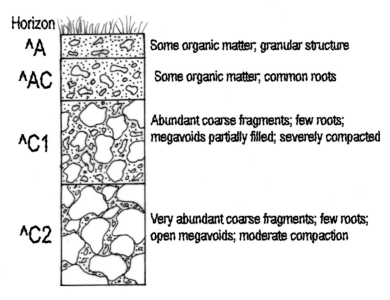

Fig. 9.7 Morphological characteristics of an eight year-old mine soil in southwestern Virginia. After Haering et al. (1993)

crusting. Compaction and surface crusting often produce mine soils with impeded drainage characterized by surface ponding, or a condition of episaturation.

A poor correlation between soil age and measured levels of organic carbon has frequently been reported for mine soils, but this was probably caused by methodological problems which made it difficult to distinguish between carbon in coal versus soil organic matter. There is little doubt that once a good vegetal cover is established, levels of soil organic matter can build up rapidly based on observations that thin A horizons develop in the upper 10–15 cm of a mine soil profile in as little as 2–4 years (Shafer et al. 1980; Daniels and Amos 1985; Ciolkosz et al. 1985; Haering et al. 1993, 2004). CEC is initially very low, but increases over time as soil organic matter accumulates (Table 9.2). Natural revegetation of mine soils can be retarded by a lack of plant-available N, low levels of microbial activity, or by a very low pH and the presence of toxic levels of metals generated by acid mine drainage.

Accelerated chemical weathering has been widely reported during the early stages of coal mine soil formation. This has been attributed to rapid leaching associated with the presence initially of large macropores in loosely packed spoils, and to the large surface area of unweathered, abraded mineral surfaces which are in disequilibrium with surface environmental conditions. Unfortunately, there is often a poor correlation between chemical properties and mine soil age due to the inherent variability in mine soil properties. For example, it is not uncommon to find very acid soils (pH 3.5–4.4) within a few meters of moderately basic (pH 7.9–8.4) soils (Haering et al. 2004). During the early stages of weathering, mine soils can be higher in pH, exchangeable bases, water-soluble Fe, and total P than local natural

9.3 Coal Mine-Related Anthrosoils

Table 9.2 Comparison of properties of barren versus well vegetated five year-old mine soils at adjacent sites in southwestern Virginia (Daniels and Amos 1985), and a 25 year-old mine soil in West Virginia (Thurman and Sencindiver 1986)

Horizon	Depth (cm)	pH	Organic C (%)	CEC (cmol kg^{-1})	HCO$_3$-ext. P (mg kg^{-1})	Sand (%)	Silt (%)	Clay (%)	Coarse fragments (%)	Bulk density g cm^{-3}
5 year old mine soil (barren, Virginia)
^A1	0–10	5.3	0.41	3.10	1.5	74	19	7	36	1.5
^AC	10–23	4.9	0.37	3.34	1.5	71	15	14	38	1.6
^C	23–100+	5.2	0.49	3.93	0.0	61	26	13	31	1.8
5 year old vegetated mine soil (well vegetated, Virginia)
^A1	0–10	5.8	0.28	4.03	4.5	72	20	8	43	1.4
^A2	10–32	5.1	0.12	4.05	4.5	76	17	7	55	1.5
^C1	32–52	5.9	0.12	4.17	3.0	76	17	8	50	1.5
^C2	52–100+	6.1	0.20	3.88	3.0	77	16	7	62	1.6
25 year old mine soil (West Virginia)
^A	0–12	5.0	1.9	–	–	20	50	30	41	1.8
^C1	12–42	5.1	1.8	–	–	26	47	27	44	1.9
^C2	42–77	4.9	2.1	–	–	26	48	26	45	1.8
^C3	77–100	4.8	1.9	–	–	30	44	26	45	1.8

Horizons modified from original decriptions

soils. Young mine soils are also frequently very high in soluble salts, particularly gypsum, derived either from sulfate-rich roof rocks, or by reaction of acid mine drainage and calcareous rocks according to Eq. 9.5. Accumulation of salt occurs rapidly, within 2–3 years or less, and is reflected in elevated values of electrical conductivity (>5000–8000 $\mu S\ cm^{-1}$). Leaching of gypsum can occur in <30 years, as shown by changes in soluble salt content as a function of profile depth (Fig. 9.8). Overall, the chemical properties of mine soils show a strong correlation with rock type, but exhibit certain properties which also seem to evolve with increasing mine soil age.

Levels of microbial biomass are initially very low, and tend to be limited by the low levels of soil organic matter generally present in young mine soils (Machulla et al. 2005; Ingram et al. 2005). Naprasnikova (2008) studied a chronosequence of mine soils in the Siberian region of Russia. Levels of microbial biomass, organic carbon, nitrogen and phosphorous were initially all very low, but increased over time such that levels were approaching those of natural soils after about 10 years. Dunger et al. (2001) studied the succession of soil organisms in German mine soils that were up to 46 years old. They found that nematodes, earthworms and various microarthropods were able to rebound to levels similar to those in natural forest soils within about 10–20 years after replanting with deciduous trees. They found that both amount and type of organic matter determined the success of soil fauna re-establishment.

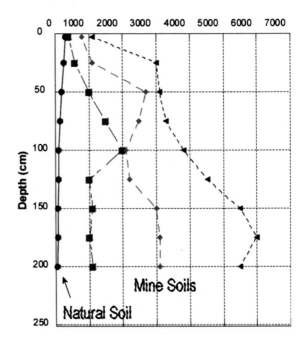

Fig. 9.8 Soluble salt (electrical conductivity) profiles for natural soil and mine soils in Pennsylvania. Soluble salts are rapidly leached from upper parts of mine soils and accumulated in subsoil within 2–7 years. After Ciolkosz et al. (1985)

Sulfuric horizons have not been widely reported in mine soils from the United States, apparently because of legal requirements that potentially toxic carbonaceous spoil materials be buried beneath non-carbonaceous spoil materials. Bini and Gaballo (2006) examined anthropogenic soils in Italy formed in spoils derived from a sulfide-rich skarn mined for Cu, Pb and Zn. They found that 35 year-old sulfidic mine soils were characterized by immature A–C soil profiles. Some mine soil profiles that were up to 100 cm thick did contain cambic horizons and were classified as Inceptisols. However, mine soils thought to be as much as 2700 years old showed relatively little morphological development. Pedogenesis was apparently greatly stunted by the long-term effects of acid mine drainage, and the presence of high levels of phototoxic metals.

Several workers proposed amending U.S. Soil Taxonomy in order to classify mine soils as a Spolent suborder of Entisol (e.g., Thurman and Sencindiver 1986), and offered suggestions for classifying mine soils down to the family level. Alternatively, it was proposed that a Spolic subgroup of Udorthent be added to accommodate mine soils (Fanning and Fanning 1989). More than 30 soil series were established to accommodate Appalachian mine soils, which were all classified as Entisols (Typic Udorthents) according to Soil Taxonomy (Sencindiver and Ammons 2000; Haering et al. 2005). Mine soils are probably classified mainly as an Anthrodensic or Anthroportic subgroup of Entisol or Inceptisol, according to the current U.S. system of Soil Taxonomy (Soil Survey Staff 2014).

Mine soils are classified as Technosols, according to the World Reference Base (IUSS Working Group 2015), based on their origin as a result of human technology. The WRB defines an artifact as something in the soil recognizably made or extracted from the Earth by humans. Thus, mine soils are Spolic Technosols, defined as having a layer ≥ 20 cm thick within 100 cm of the surface with $\geq 20\%$ (by volume) artifacts in the form of mine spoil. Mine soils are classified as Technogenic superficial formations in the Russian soil classification system, and as Anthroposols in the Australian system.

References

Akeil A, Koldas S (2006) Acid mine drainage (AMD): causes, treatment and case studies. J Clean Prod 14:1139–1145

Sencindiver JC, Ammons, JT (2000) Minesoil genesis and classification. In: Barnhisel RI (ed) Reclamation of drastically disturbed land. Madison, WI, pp 595–613 (Agron Monog 41, SSSA)

Bini C, Gaballo S (2006) Pedogenic trends in anthrosols developed in sulfidic mine spoils: a case in the Temperino mine archaeological area (Campiglia Marittima, Italy). Quat Intern 156–157:70–78

Chang-sheng J, Zhao-xue C, Qing-hua C (2009) Surface coal-mining practice in China. Procedia Earth Planet Sci 1:76–80

Ciolkosz EJ, Cronce RC, Cunningham RL, Peterson GW (1985) Characteristics, genesis and classification of Pennsylvania minesoils. Soil Sci 139:232–238

Daniels WL, Amos DF (1985) Generating productive topsoil substitutes from hard rock overburden in the southern Appalachians. Environ Geochem Health 7:8–15

Diessel CFK (1992) Coal-bearing depositional sequences. Springer, Berlin, Germany, 721 pp

Dunger W, Wanner M, Hauser H, Hohberg K, Schulz HJ, Schwalbe T, Seifert B, Vogel J, Voigtlander K, Zimdars B, Zulka KP (2001) Development of soil fauna at mine sites during 46 years of afforestation. Pedobiologia 45:243–271

Fanning DS, Fanning MCB (1989) Soil morphology, genesis, and classification. Wiley, New York, 395 pp

Grim EC, Hill RD (1974) Environmental protection in surface mining of coal. U.S. Environmental Protection Agency, Office of Research and Development, EPA-670/2-74-093, 277 pp

Haering KC, Daniels WL, Roberts JA (1993) Changes in mine soil properties resulting from overburden weathering. J Environ Qual 22:194–200

Haering KC, Daniels WL, Galbraith JM (2004) Appalachian mine soil morphology and properties: effects of weathering and mining method. Soil Sci Soc Am J 68:1315–1325

Haering KC, Daniels WL, Galbraith JM (2005) Mapping and classification of southwest Virginia mine soils. Soil Sci Soc Am J 69:463–472

Holz M, Kalkreuth W, Banerjee I (2002) Sequence stratigraphy of paralic coal-bearing strata: an overview. Intern J Coal Geol 48:147–179

Ingram LJ, Schuman GE, Stahl PD, Spackman LK (2005) Microbial respiration and organic carbon indicate nutrient cycling recovery in reclaimed soils. Soil Sci Soc Am J 69:1737–1745

IUSS (International Union of Soil Scientists) Working Group (2015) World Reference Base for Soil Resources 2014 (update 2015). International soil classification system for naming soils and creating legends for soil maps. World Soil Resources Reports No. 106. FAO, Rome, 193 pp

Kittrick JA, Fanning DS, Hossner LR (1982) Acid sulfate weathering. Soil Sci Soc Am Spec Publ 10:234 pp (Madison, WI)

Machulla G, Bruns MA, Scow KM (2005) Microbial properties of mine spoil materials in the initial stages of soil development. Soil Sci Soc Am J 69:1069–1077

Naprasnikova EV (2008) Biological properties of soils on mine tips. Eurasian Soil Sci 41:1314–1320

Shafer WM, Nielsen GA, Nettleton WD (1980) Minesoil genesis in a spoil chronosequence in Montana. Soil Sci Soc Am J 44:802–807

Soil Survey Staff (2014) Keys to soil taxonomy, 12th edn. U.S. Department of Agriculture, Natural Resources Conservation Service, 372 pp

Thomas L (2012) Coal geology, 2nd edn. Wiley-Blackwell, Chichester, Sussex, 454 pp

Thurman NC, Sencindiver JC (1986) Properties, classification and interpretations of minesoils at two sites in West Virginia. Soil Sci Soc Am J 50:181–185

Toy TJ, Hadley RF (1987) Geomorphology and reclamation of disturbed lands. Academic Press, 480 pp

Chapter 10
Anthropogenic Soils in Urban Settings

Anthropogenic soils found in urban areas vary as a function of land use and city structure. They are classified primarily as Technosols, the central concept of which is a soil formed in HTM characterized by an abundance of technogenic artifacts. Urban soils in residential areas are Urbic Technosols characterized by artifacts indicative of human habitation, whereas those associated with industrial land are Spolic Technosols with an artifact assemblage comprised of industrial wastes. Urban soils typically have an ^Au-^Cu profile, although cambic-like horizons may form in 18–70 years. They are characteristically heterogeneous and often suffer from excessive compaction, excessive artifact content, diminished biological activity, and reduced infiltration due to surface crusting or water repellency. They generally have elevated levels of pH, exchangeable bases, and carbonate content, but levels of organic C, N and P tend to be very low in recently deposited HTM. Melanization results in the formation of ^A horizons in as little as 12 years under an udic moisture regime with a cover of grass. Under such conditions, urban soils are resilient and may re-establish properties similar to those of natural soils in 30–100 years.

10.1 Introduction

Anthropogenic soils in urban settings are largely members of the Technosol Reference Soil Group in the World Reference Base. The characteristic feature of a Technosol is the presence of materials manufactured, altered, or exposed by human technology that otherwise would not occur at the Earth's surface including pavement, artifacts, building rubble, dredging and mine spoils, etc. Different types of Technosol are distinguished further using qualifiers indicating whether the artifacts are primarily rubble and refuse of human settlements (Urbic Technosols), industrial wastes (Spolic Technosols), organic wastes (Garbic Technosols), and so forth. They are generally referred to as "urban soils" which, broadly defined, includes all soils mapped in an urban setting. It is widely recognized that there is a close correlation between urban soil type and land use category, hence city structure generally controls the nature and extent of urban soils. City structure refers to the spatial arrangement of land use in an urbanized area. It explains why and how landscapes were modified by human activity, and thus accounts for the general geospatial (geographic) distribution of urban anthropogenic soils. City structure varies according to historic settlement pattern and geocultural setting, thus North American cities tend to differ in structure from those in Europe, Asia and South America.

Three contrasting models were proposed during the early 20th century to explain the structure of North American cities. These are known as the "classic models." In the *concentric zone model* (based on Chicago, Illinois) of Burgess (1924), cities grow outward from a central point in the form of concentric bands (Fig. 10.1a). The innermost band represents the central business district (CBD), which is surrounded by an industrial zone with poor quality housing. Outward from this zone are rings in which residential land contains progressively better quality housing. Burgess thought that the demand for better housing drove more affluent residents away from the aged, deteriorated housing near the industrialized city center. The *sector model* of Hoyt (1939) involves the development of mixed land uses in the CBD, after which the city expands outward along pie-shaped sectors following major railroad lines (Fig. 10.1b). The *multiple nuclei model* of Harris and Ullman (1945) postulated that cities do not necessarily grow around a CBD, but rather around various city centers which develop into independent areas because of their specialized activities (Fig. 10.1c). Although elements of the classic models can be seen in many cities worldwide, additional geocultural factors operated during the latter part of the 20th century to shape the structure of modern cities. For example, Vance's (1964a, b, 1990) *urban realms model* extended the multiple nuclei model to take the effects of terrain and the automobile into account (Fig. 10.1d). This model worked well for San Francisco and Los Angeles, California.

Most major cities in the United States are characterized by a CBD. This is usually the part of the city which developed first, and then became the focus of industrial and governmental activity during the Industrial Revolution. The CBD is often located at the geographic city center, comprises the inner city, and serves as a hub for train, subway and bus transportation. Today, the CBD is typically a region of manufactured land built on thick deposits of human-transported material from the 18th and 19th century. The CBD is typically ringed by a middle zone comprised of mixed industrial and residential land which has experienced multiple demolition cycles, and where soils developed in human-transported material were strongly affected by coal-related activities fueling the Industrial Revolution. This middle zone is transitional to an outer zone of peripheral urban and suburban residential and commercial land. Urban soils in the middle and outer zones are generally less disturbed than those closer to the city center. This urban structural pattern is partly the result of the proliferation of the automobile and the development of the

10.1 Introduction

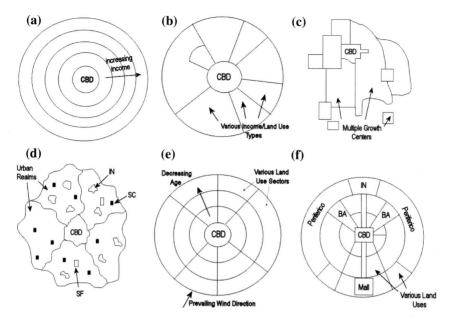

Fig. 10.1 Comparison of city structure models. **a** Burgess (1924) concentric ring model. **b** Hoyt (1939) sector model. **c** Multiple nuclei model of Harris and Ullman (1945). **d** Urban realms model of Vance (1990). **e** Mann's model (Britain). **f** Griffin-Ford model (Ford 1996). *CBD* central business district; *IN* industrial center; *SC* shopping center; *SF* major sports facility; *BA* barrios

interstate highway system, which allowed more affluent residents to move to the suburbs, or beyond. Thus, a commuter-based society has developed in which drivers commute to and from the CBD or, more often, to outer city centers and suburban industrial parks. This has led to much inner city deterioration and blight. The construction of interstate highways which ring and bypass the inner city, such as the beltways of Baltimore, Maryland and Washington, D.C., also contributed to inner city decay. In some places (e.g., Chicago and New York), the CBD remained vibrant throughout the 20th century. Elsewhere, a once blighted inner city has been largely redeveloped and revitalized via gentrification (e.g., Baltimore), or is still struggling to do so (e.g., Detroit).

In North America, particularly in the United States, real estate developers often had more political clout than urban planners, hence cities there typically underwent very rapid, uncontrolled post-WWII growth in the form of unplanned, urban sprawl. In contrast, European cities controlled post-WWII development and prevented cities from merging into conurbations by creating greenbelts that were protected from urbanization. European cities are also usually much older than those in North America and are often characterized by a prominent historic feature in the city center, such as a cathedral or castle, which attracts tourists and stimulates gentrification. Thus, the European city center may be vibrant and possibly inhabited by upper class residents. Elements of the Burgess and Hoyt models can be seen in some British cities, and London perhaps has elements of all three classic models. In 1965, Mann proposed a model for some English cities which involved both concentric rings and sectors, but suggested that because of its long history of severe air pollution, prevailing wind direction strongly influenced spatial pattern of development (Fig. 10.1e). Lawton's model from 1973 suggested that British cities did not always conform to the concentric ring structure because the development of cheap council housing on the periphery altered the ring pattern. He suggested that Liverpool has a ring structure, but it reflects a cyclic history of prosperity rather than an outward migration by more affluent residents. The construction of ring roads, and economic factors similar to those in the U.S. have also led to the modern structure of British cities. For example, a decline in heavy industry, the rise in franchises, and the growth of planned outer cities or edge cities has led to deterioration of the CBD. British cities have undergone stages of development similar to those in the United States, although the Industrial Revolution began many decades sooner there: (1) a phase of urbanization which attracted rural immigrants to the CBD, (2) a later phase of suburbanization during which the wealthy migrated to the suburbs resulting in a decline of the CBD and shipping port-related activities, (3) the current phase of "counter-urbanization" as the wealthy move beyond the suburbs. Cities in Latin America and Asia also contrast with those in North America (Ford 1996). Their structures show a hybrid pattern of concentric zones and sectors (Fig. 10.1f), but they may lack a CBD and have a separate market center or zone. They generally show a decrease in wealth with increasing distance from the city center, culminating in the periferico (barrio).

10.2 Urban Land Classification

Land cover and land use are generally closely related (Anderson et al. 1976), and there is usually a strong correlation between land cover and such features as population density, urban runoff, soil type, contamination, etc. (Arnold and Gibbons 1996; Cuffney et al. 2008; Hazelton and Murphy 2011). Thus, in this book, urban land is classified into five basic categories (Table 10.1): (1) Greenspace, (2) Cemeteries, (3) Residential land, (4) Industrial land, and (5) Manufactured (sealed) land. **Greenspace** is predominantly natural, undisturbed land, but some landscaping or construction-related disturbance may have occurred. It is also possible to have former demolition sites, landfills or perhaps reclaimed industrial sites which have been repurposed as greenspace. Although natural soils tend to predominate in such settings, it is not uncommon for soils in green areas to show evidence of human disturbance. **Cemeteries** have gravesites characterized by Anthrosols formed in human-altered material, as discussed in Chap. 8. **Residential land** is typically a complex mosaic of paved land, vacant land produced by building

10.2 Urban Land Classification

Table 10.1 Classification and characteristics of urban land

Urban land type	Description	Urban soil types	Artifacts Abundance	Type[a]
Greenspace	Parkland, historic sites and monuments, golf courses, urban agriculture; rare artifacts and evidence of human disturbance	Natural soils predominant; possible anthroportic soils	None to sparse	–
Cemeteries	Gravesites, crypts and mausoleum; ornamental landscaping, stockpiled soil	Anthraltic and anthroportic soils predominant; natural soils and anthropogenic phases of natural soils possible	Common in grave soils	–
Residential	Paved land, single-multifamily dwellings, schools, light commercial (service industry) buildings; monocyclic and polycyclic demolition sites	Sealed soils, anthropic and anthroportic construction-related and demolition site soils predominant; anthropogenic phases of natural soil series common	Few to moderately abundant	b, m, t, d, c, w, h, g, n
Industrial	Heavy commercial and manufacturing, power plants, airports, highways, railroads, landfills, mined areas, dredged land	Anthroportic soils predominant; sealed soils	Moderately to very abundant	c, k, d, s, b, m, g
Manufactured land	Ground surface covered with impervious materials (asphalt, concrete, etc.)	Sealed soils predominant	None	–

[a]Artifacts: *b* brick; *m* mortar; *t* concrete; *d* coal cinders; *c* coal; *k* coked coal; *w* wood; *h* charcoal; *g* glass; *n* nails; *s* iron smelting slag

demolition, and land on which extant buildings may range from occupied and well maintained, to abandoned and derelict. Vacant land may represent one cycle of demolition (monocyclic), or many cycles (multicyclic). Anthropogenic soils associated with residential land are generally characterized by rich assemblages of artifacts indicative of human habitation. **Industrial land** includes sites of heavy industry, land modified by mining and dredging operations, and land utilized for solid waste disposal (landfills) and transportation-related activities (highways, railroads, airports). It is typically severely disturbed, and characterized by a wide variety of artifact-rich soils which are often highly contaminated. Buildings may have been demolished leaving behind a great volume of debris and many underground structures (e.g., drains, pipelines, storage tanks) which may be more or less buried beneath fill. **Manufactured land** is that which is completely or almost

Fig. 10.2 Reconnaissance soil survey of New York City (Soil Survey Staff 2005). Map units 1–4 comprising most of Manhattan Island are manufactured land

completely covered by impervious surfaces such as paved streets, sidewalks and buildings. For example, almost the entire area of Manhattan Island in New York City was mapped as manufactured land (Fig. 10.2).

10.3 General Characteristics of Urban Soils

Cemetery Anthrosols were described previously in Chap. 8, and given that urban greenspace is occupied essentially by natural soils, these types of landscapes are not discussed further here. Eleven different types of Technosol are recognized in the World Reference Base (Table 10.2), which may potentially be found associated with residential, industrial and manufactured land. Landfill soils were described elsewhere by Meuser (2010). Hence, the focus here is on Urbic and Spolic Technosols associated with residential and industrial lands, respectively. These soils have formed from human-transported material (HTM), as defined in Soil Taxonomy (Soil Survey Staff 2014). The few known studies of soils sealed beneath technic hard material (manufactured land) will also be discussed.

The main characteristics of anthropogenic urban soils formed in HTM (Table 10.3) are: (1) simple, weakly developed profile morphology, (2) truncated, stratified and buried soil profiles, (3) extensive geospatial variability, (4) loss of aggregation and soil structure, (5) irregular organic matter distribution, (6) excessive compaction, (7) excessive artifact content, (8) altered soil chemical properties,

10.3 General Characteristics of Urban Soils

Table 10.2 Qualifiers used to distinguish different types of Technosol potentially found in urban settings, according to the World Reference Base (IUSS Working Group 2015)

Qualifier	Description
Urbic	Layer ≥ 20 cm thick within ≤ 100 cm of surface containing significant amount of rubble and refuse of human settlements
Spolic	Layer ≥ 20 cm thick within ≤ 100 cm of surface containing significant amount of industrial wastes: mine spoils, dredging, slag, ash, rubble, etc.
Hyperskeletic	Layer containing $\leq 20\%$ fine earth fraction within ≤ 75 cm of surface
Garbic	Layer ≥ 20 cm thick within ≤ 100 cm of surface containing significant amount of organic wastes; typically associated with landfills
Reductic	Reducing conditions caused by methane or carbon dioxide; typically associated with landfills
Linic	Geomembrane within ≤ 100 cm of surface
Ekranic	Layer of technic hard material within ≤ 5 cm of surface
Leptic	Layer of technic hard material within ≤ 100 cm of surface
Isolatic	Soils found on rooftops or in pots
Subaquatic	Permanently submerged by water ≤ 200 cm deep
Cryic	Having permafrost within ≤ 100 cm of surface

Table 10.3 General characteristics of urban soils formed in human-transported material (HTM)

Urban soil characteristic	Description	Implications
Soil morphology	Artifact-rich, simple ^Cu, or ^Au-^Cu, profile common; ^Au horizon may develop in 12–24 years; possible cambic-like ^Bw horizon in 18–70 years	Rapid rebound to natural conditions in 30–100 years facilitates repurposing of vacant urban land (e.g., urban horticulture)
HTM stratigraphy	Truncated (loss of A horizon), stratified and buried soil profiles common; lithologic discontinuities common due to excavation or grading with earthmoving equipment	Difficulty with soil mapping and classification
Extensive geospatial variability	Extensive horizontal variation in profile characteristics due to heterogeneous nature of fill, and excavation or grading with earthmoving equipment	Difficulty with soil mapping and classification
Loss of aggregation and soil structure	Massive or single grain structure common due to excavation and backfilling	Impeded aeration and drainage inhibits plant growth and microbial activity
Irregular organic matter distribution	Highly variable vertical and lateral organic C content as a result of excavation and backfilling	Complicates soil classification

(continued)

Table 10.3 (continued)

Urban soil characteristic	Description	Implications
Excessive compaction	Densic contact and densic materials common as a result of earthmoving equipment	Limited infiltration, root penetration, and aeration, inhibits plant growth and microbial activity; tree mortality common
Excessive artifact content	Incorporation of artifacts from human habitation and demolition debris during backfilling	Ground subsidence as non-persistent artifacts decompose; persistent artifacts make soil impervious and difficult to map with soil auger; limits root penetration and difficult to till by urban agronomists
Altered soil chemical properties	Elevated levels of pH, exchangeable bases, carbonate content common; levels of organic C, N and P may be diminished	Possibly more fertile than natural soils, especially after artificial additions of organic matter
Altered biological activity	Soil rendered biologically sterile and prone to colonization by invasive species	Rapid rebound to natural conditions in 30–100 years facilitates repurposing of vacant urban land (e.g., urban horticulture)
Altered geophysical properties	Elevated values of pH, electrical conductivity, magnetic susceptibility and penetrability common	Facilitates use of geophysical methods for mapping urban soils
Modified soil temperature and moisture regimes	Generally warmer and drier than natural soils	Contributes to accelerated organic matter decomposition and weathering of inorganic artifacts
Surface crusting	Physical sealing and water repellency common in silty and sandy soils, respectively	Limited infiltration leading to ponding or increased runoff and erosion; restricted seedling emergence

(9) diminished biological activity, (10) altered geophysical properties, (11) modified soil temperature and moisture regimes, and (12) surface crusting.

^A horizons have been widely reported to have developed in urban soils within a few decades. For example, urban soils in Detroit, Michigan USA formed in recently deposited HTM less than 5–10 years old were characterized by an ^Cu profile, whereas ^A-^C profiles were present after 10–25 years of pedogenesis (Howard et al. 2013). Howard and Olszewska (2011) described an ^Au horizon 16 cm thick containing 1.2% organic C in a demolition site soil formed under a cover of grass in an udic moisture regime after only 12 years of pedogenesis. However, the distribution of cambic or cambic-like horizons is highly irregular. They have been reported to form in some urban soils on a time scale of decades, but are lacking in other situations after many decades of pedogenesis. Howard and Shuster (2015)

10.3 General Characteristics of Urban Soils

described a cambic-like ^Bwgu horizon in a somewhat poorly drained 18 year-old demolition site soil based on the presence of structure. Cambic and cambic-like horizons were also present in some well drained urban soils in the same area, but absent in others. The morphological characteristics of these anthropogenic soils in Detroit, Michigan are compared with those of a natural soil in Table 10.4. Detroit was built on a late Pleistocene glacial lakebed, and the natural terrain is characterized by Typic Hapludalfs on well-drained beach ridges, and Typic Hapludolls and Argiudolls (Fig. 10.3a) on poorly drained lakebed plains (mollic epipedons formed as a result of wetness). The anthropogenic soils were all formed in deposits of HTM overlying buried natural soils (Fig. 10.3b–d). The 68 year-old demolition site soil was characterized by an ^Au-^Cu profile, whereas the 70 and 85 year-old soils had ^A-^Bw profiles (Table 10.4). The cambic-like ^Bw horizons were identified on the basis of weak structure, which occupied a limited volume of soil, hence they do not meet the full requirements for a cambic horizon. The ^Au horizons are black (10YR2/1) because they contain significant amounts of coal-related wastes (sand-sized coal and carbonaceous shale, flyash, soot, etc.) which are a legacy from the coal-burning era of the late 19th and early 20th centuries. These ^Au horizons visibly resemble mollic epipedons in the field. Mottling is common elsewhere as a result of restricted drainage caused by compaction (e.g., Howard and Olszewska 2011). Mottling is typically in the form of redoximorphic depletions (value ≥ 4 and chroma ≤ 2). The texture of urban anthrosoils is often gravelly (from abundant artifacts) and may be sandier than natural soils locally (Craul 1985; Burghardt 1994). The structure of ^Au horizons typically ranges from massive or single grain, to weak, fine granular; weak, fine angular blocky; or platy (densic materials).

Lithologic discontinuities and various types of irregular stratification, produced by human transport and deposition during the backfilling process, are characteristic features of urban soils (Fig. 10.4). They contribute to their extensive geospatial and stratigraphic variability (Fig. 10.5). In a study of 100 urban soil profiles in Washington, D.C., 95% were found to have at least one lithologic discontinuity, and several had as many as five (Short et al. 1986a). Horizon boundaries are usually abrupt, wavy or irregular, and occasionally broken, as a result of the backfilling process. ^Au and ^Cu horizons may show evidence of decalcification and calcification, respectively, evident in the field by reaction to 10% HCl. Petrographic evidence suggests that this is due locally to the leaching of carbonate from calcareous artifacts (Howard and Olszewska 2011; Howard et al. 2013). However, argillans or other indications of illuviation of pedogenic clay are not known from urban anthrosoils. Krotovina representing earthworm burrows can be present after only a few years, and abundant in urban soils older than about 30 years (Howard et al. 2013, 2015). In the situation where a pre-existing natural soil experiences relatively little disturbance, e.g., as a result of minimal construction-related grading and site preparation, the topsoil may retain some of its original characteristics and may only suffer from mixing and a loss of aggregation. Soil materials with massive or single grain structure can also become mixed with aggregated topsoil, or subsoil containing structure, during backfilling. This inherited structure can appear as a

Table 10.4 Morphological characteristics of a natural soil compared with anthropogenic soils at urban residential and industrial sites in Detroit, Michigan (NRCS Soil Survey Staff 2016, unpub. data)

Horizon	Depth (cm)	Description
Pewamo series (natural soil: Typic Argiudoll); Wayne County, Michigan		
Ap	0–28	Very dark grayish-brown (10YR3/2, moist) sandy loam; weak, moderate fine granular to single grain; moist friable, non-sticky, non-plastic; common fine roots; no effervescence; gradual smooth boundary
Bt	28–38	Strong brown (7.5YR4/6) sandy loam; very weak, coarse subangular blocky; moist friable, non-sticky, non-plastic; common thin discontinuous clay films; few fine to medium roots; common medium krotovina; no effervescence; clear smooth boundary
BC	38–55	Dark yellowish-brown (10YR4/6) loam sand; massive to single grain; moist friable, non-sticky, non-plastic; few fine roots; few medium krotovina; no effervescence; gradual smooth boundary
Cg	55–100	Strong brown (10YR6/5) fine sand with common, medium, prominent yellowish-red (5YR4/6) mottles; massive to single grain; few thin discontinuous clay films; moist loose, non-sticky, non-plastic; very few fine roots; weak effervescence in lower part; stratified, overlying clayey diamicton
Undemolished building site (Oxyaquic Hapludoll, ~85 years old); Detroit, Michigan		
^Au	0–25 cm	Black (10YR2/1) loamy fine sand, dark gray (10YR4/1), dry; weak medium subangular blocky parts to weak fine granular structure; very friable; few medium roots throughout and common fine roots throughout; 7% nonflat subangular indurated 2–20-mm mixed rock fragments; 1% irregular subangular 20–74 mm brick; slightly alkaline, pH 7.4; abrupt smooth boundary
^Bwu	25–66	Dark yellowish-brown (10YR3/4) sand; weak medium subangular blocky parts to single grain; very friable; common fine roots throughout; 2% nonflat subangular indurated 2–20-mm mixed rock fragments and 10% nonflat subrounded weakly cemented 20–75-mm iron-manganese concretions; 1% irregular subangular 75–249 mm metal and 1% irregular subangular 75–249 mm brick; 10% krotovinas (volume percent); slightly alkaline, pH 7.7; abrupt smooth boundary
Ab	66–76	Black (10YR2/1) loamy fine sand; weak medium granular structure; friable; few medium roots throughout and few fine roots throughout; 1% nonflat subangular indurated 2–20-mm mixed rock fragments; 2% flat angular 20–74 mm glass; slightly alkaline, pH 7.6; clear wavy boundary
Bhsb	76–86	Dark reddish-brown (5YR3/2) loamy sand; weak coarse subangular blocky structure; friable; common medium roots throughout and few fine roots throughout and few coarse roots throughout; 1% nonflat subangular indurated 2–5-mm mixed rock fragments; slightly alkaline, pH 7.4; clear wavy boundary
Bwb	86–117	Yellowish-red (5YR4/6) sand; weak medium subangular blocky parts to single grain; loose, weakly cemented; few fine roots throughout; 10% coarse prominent (5YR4/4) masses of oxidized iron in matrix and 40% coarse prominent weakly cemented (7.5YR4/6) masses of oxidized iron in matrix; neutral, pH 7.2; gradual smooth boundary
C	117–142	Yellowish-red (5YR5/6) sand; weak thin platy structure; loose; few fine roots throughout; 10% medium distinct (10YR5/2) iron depletions and 30% coarse prominent weakly cemented (7.5YR4/6) masses of oxidized iron in matrix; neutral, pH 7.2; gradual smooth boundary

(continued)

10.3 General Characteristics of Urban Soils

Table 10.4 (continued)

Horizon	Depth (cm)	Description

Demolition site soil (Anthroportic Udorthent, ~68 years old); Detroit, Michigan

Horizon	Depth (cm)	Description
^Au	0–15	Very dark grayish-brown (10YR3/2) to black (10YR2/1) sandy loam, grayish brown (10YR5/2), dry; weak medium subangular blocky parts to moderate medium granular structure; friable; common very fine roots throughout and common fine roots throughout; 13% gravel; 5% gravel size bitumen (asphalt); strong effervescence; moderately alkaline, (pH 7.9); abrupt smooth boundary
^Cu	15–56	65% brown (10YR4/3) and 35% very dark gray (10YR3/1) gravelly artifactual clay loam; massive; firm; common fine roots throughout; common medium distinct yellowish brown (10YR5/6) masses of oxidized iron in matrix and common coarse distinct moderately cemented yellowish-red (5YR4/6) iron-manganese concretions; 9% gravel; 4% gravel size bitumen (asphalt); 8% cobble size bitumen (asphalt); 10% gravel size brick; 10% cobble size brick; common very dark gray (10YR3/1) krotovinas; strong effervescence; moderately alkaline, (pH 8.2); abrupt wavy boundary
2Ab	56–102	Gray (2.5Y5/1) sandy clay loam; strong medium subangular blocky structure; firm; common fine roots between faces of peds; common coarse prominent yellowish-brown (10YR5/6) and (10YR5/8) masses of oxidized iron on faces of peds; 2% gravel; common very dark gray (10YR3/1) krotovinas; moderately alkaline, (pH 8.3); gradual wavy boundary
2C	102–150	Yellowish-brown (10YR5/4) clay loam; massive; firm; common fine roots in cracks; common coarse prominent yellowish-brown (10YR5/8) masses of oxidized iron in matrix and common medium distinct gray (5Y5/1) iron depletions in matrix; 5% gravel; strong effervescence; moderately alkaline, (pH 8.3); abrupt wavy boundary
Cd	150–203	Yellowish brown (10YR5/4) loam; moderate thick platy structure

Industrial (Train Yard) site soil (Anthroportic Udorthent, ~70 years old); Detroit, Michigan

Horizon	Depth (cm)	Description
^Au1	0–10	Very dark gray (10YR3/1) loamy sand; weak fine granular; very friable, non-sticky, non-plastic; many fine roots; contains 3% artifacts (coal cinders); no weak reaction; abrupt wavy boundary
2^Au2	10–20	Black (10YR2/1) sandy gravel; single grain; loose; 70% artifacts (coal cinders, mortar, plasterboard) strong reaction; few medium roots; abrupt wavy boundary
3^Bwu	20–43	Dark yellowish-brown (10YR4/4) loamy sand; weak, medium subangular blocky structure; friable; 5% artifacts (corroded railroad spikes); few fine roots; abrupt broken boundary
3^Bw	43–58	Yellowish-brown (10YR5/6) sand; weak, thin, platy structure; very friable; 1% artifacts (cinders); few fine roots; abrupt smooth boundary
3Ab	58–71	Black (10YR2/1) sand; weak fine granular structure; very friable; few fine to coarse roots; abrupt wavy boundary
3Bgb	71–84	Grayish-brown (2.5Y5/2) sand with common, fine, prominent, strong brown (7.5YR4/6) Fe-oxide concentrations along root channels; weak medium platy structure; friable; few fine to medium roots; clear wavy boundary

Fig. 10.3 Comparison of natural versus anthropogenic soils. **a** Natural soil (Typic Argiudoll). **b** Undemolished building site soil (Oxyaquic Hapludoll). **c** Demolition site soil (Anthroportic Udorthent). **d** Industrial (Train Yard) site soil (Anthroportic Udorthent). Scale in cm. Photos by J. Howard and USDA-NRCS

relict feature giving the false impression that an anthropogenic soil forming in recently deposited HTM is in an anomalously advanced pedogenic stage of development. Variable amounts of topsoil are often incorporated into HTM, hence

10.3 General Characteristics of Urban Soils

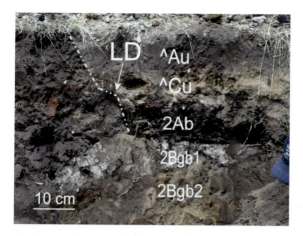

Fig. 10.4 Demolition site soil showing lithologic discontinuity (*LD*) separating two distinctly different anthropogenic soil profiles. Photo by Samantha Lynch

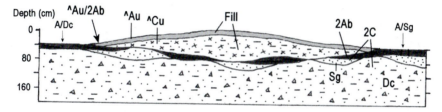

Fig. 10.5 Interpretive model of urban soil landscape in downtown Detroit, Michigan USA. A horizon is at the surface locally, or buried (*Ab*) beneath fill and partially eroded. Parent material varies from clayey diamicton (*Dc*) to gravelly lacustrine sand (*Sg*). From Howard et al. (2013)

there is frequently an irregular distribution of soil organic matter with increasing depth (Short et al. 1986a, b).

Excessive compaction is often caused by earthmoving equipment, and may be done intentionally to prepare a demolition site foundation for future construction. Compaction of the topsoil can result directly from surface loading and compression, whereas vibrations created by mechanized equipment may produce compaction at greater depths (Craul 1985). In general, plants that cannot root below ~30 cm depth, suffer adverse consequences particularly during times of drought. Short et al. (1986a) documented a decrease in percentage void space in urban topsoils of Washington, D.C. as a result of excessive compaction. They reported that sandy loam, loam and silt loam surface soils had bulk densities ranging from 1.6 to 1.7 g cm^{-1} (n = 100), with bulk densities as high as 2.0 g cm^{-1} at 30 cm depth. This greatly exceeds the upper limit of bulk density (1.3–1.4 g cm^{-1}) for unimpeded plant root growth in loamy soils (Table 10.5). Bulk densities exceeding 1.6 g cm^{-1} have also reported for urban anthrosols elsewhere in the eastern (Craul

Table 10.5 Effects of bulk density and texture on plant root growth in anthropogenic urban soils (Soil Survey Staff 2000)

Soil texture	Optimal bulk density g cm^{-1}	Bulk density restricting root growth g cm^{-1}
Sand, loamy sand	<1.6	>1.8
Sandy loam, loam	<1.4	>1.8
Sandy clay loam	<1.4	>1.75
Silt, silt loam	<1.3	>1.75
Silty clay loam	<1.1	>1.65
Sandy clay, some silty clay and clay loam (>35–45% clay)	<1.1	>1.58
Clay (>45% clay)	<1.1	>1.47

1985; Pouyat et al. 2007), and western (Scharenbroch et al. 2005) United States, China (Jim 1998), and Europe (Mullins 1991). Excessive compaction often results in urban tree fatality, and reduces aeration, thereby adversely affecting soil microbial populations. Compaction can enhance preferential water flow in the vadose zone (Etana et al. 2013; Shuster et al. 2014). Preferential flow can occur along cracks and other types of macroporosity created by an abundance of artifacts. It is also well established that preferential flow can occur along the vertical burrows of earthworms such as *Lumbricus terrestris* (Edwards et al. 1992). Recent research suggests that although earthworms tend to avoid compacted soils, they can burrow through compacted layers. Preferential flow along vertical earthworm burrows in urban soils may actually be enhanced by the presence of compacted layers. This explains why groundwater water was observed in the otherwise exceptionally dry demolition site soil described in Table 10.4 which had densic layers with bulk densities of 1.9 g cm^{-3} (Fig. 10.6).

Although their texture depends on that of local surficial deposits and soils, urban anthrosoils often contain more sand and gravel than natural soils. Gravel is typically in the form of artifacts, often waste building materials, the volume of which is seen to greatly increase in urban anthropogenic soils following the beginning of the Industrial Revolution (Bridges 1991). Although artifact content is typically <10% by volume in construction-related anthrosoils, it commonly comprises 10–35% of demolition sites, and can locally be in excess of 65% by volume. A very large volume of debris was sometimes left onsite intentionally by demolition contractors to form what they call "hardcore" (Fig. 10.7), in order to prepare the site for future construction. Excessive artifact content adversely affects plant root penetration as well as infiltration and percolation rates. The presence of very sharp artifacts can also adversely affect penetration by roots and earthworms, and an abundance of biodegradable or non-persistent clasts (e.g., wood) can result in ground subsidence. The high porosity of certain types of artifacts can affect soil aeration, water-holding capacity, hydraulic conductivity, and bulk density (Burghardt 1994; Meuser 2010).

The data in Table 10.6 illustrate how the chemical properties of urban soils have been altered from those of a natural soil by human activities (also see Table 10.4).

10.3 General Characteristics of Urban Soils

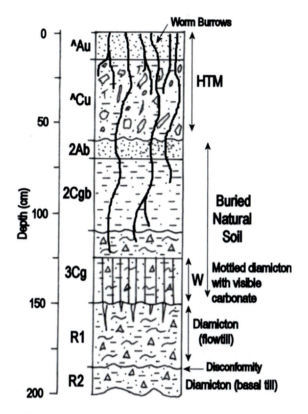

Fig. 10.6 Soil profile showing how preferential flow has occurred along vertical burrows of *Lumbricus terrestris* thus allowing vadose water to percolate through highly compacted anthropogenic subsoil (^Cu). *HTM* human-transported material; *W* groundwater

Fig. 10.7 Demolition site showing extensive accumulation of artifacts comprising "hardcore" (Detroit, Michigan USA). Scale marked in 10 cm intervals. Photos by J. Howard

Table 10.6 Physical and chemical properties of natural soil compared with anthropogenic urban soils (NRCS Soil Survey Staff, 2016; unpub. data)

Horizon	Texture	Sand %	Silt %	Clay %	Exchangeable Bases Ca cmol kg⁻¹	Mg	K	Na	Base saturation %	Org. C %	Total N %	C/N	Bray # 1 P mg kg⁻¹	pH	Bulk density g cm⁻³
Pewamo series (natural soil; Typic Argiaquoll); Wayne County, Michigan															
Ap	S	91	8	1	3.3	0.6	0.04	0	32	1.6	0.12	13	–	6.4	–
Bt	S	96	3	1	1.1	0.1	0.03	0	9	1.2	0.07	17	–	5.5	–
BC	S	97	2	1	0.7	0.1	0.02	0	6	1.0	0.06	16	–	5.3	–
Cg	S	89	10	1	0.7	0.1	0.04	0	9	0.1	0.02	7	–	5.2	–
Undemolished building site (Oxyaquic Hapludoll); Detroit, Michigan															
^Au	LS	84	10	6	8.8	1.9	Tr	–	100	4.2	0.3	15	48.4	7.4	–
^Bwu	S	92	5	4	5.8	1.1	–	–	100	1.1	0.1	17	11.0	7.7	–
Ab	LS	87	10	4	15.2	2.4	Tr	–	100	3.8	0.2	22	16.2	7.8	–
Bhsb	S	90	7	3	17.9	2.2	–	–	100	2.6	0.1	30	1.3	7.9	–
Bwb	S	94	4	2	5.2	0.7	–	–	100	0.6	0.1	10	1.5	7.8	–
C	S	97	2	1	1.4	0.3	–	–	100	0.2	–	–	3.3	7.6	–
Demolition site (Anthroportic Udorthent); Detroit, Michigan															
^Au	SL	62	26	12	38	2.4	0.3	–	100	3.1	0.2	13	20.1	7.9	1.57
^Cu	CL	32	40	28	44	4.2	0.4	–	100	1.9	0.1	13	2.9	8.2	1.88
BCgb	SCL	65	14	21	8.5	3.3	0.3	0.1	100	0.3	0.1	6	1.9	8.3	1.94
C	CL	40	32	28	46.3	6.8	0.3	Tr	100	1.9	0	5	0.2	8.3	1.92
Cd	L	33	43	24	48.3	4.3	0.2	0.1	100	3.4	0	15	0.1	8.4	–
Industrial (Train Yard) site (Anthroportic Udorthent); Detroit, Michigan															
^Au	L	51	34	15	20.2	6.0	0.5	Tr	100	6.6	0.4	16	11.2	7.3	–
2^Au	LS	76	20	4	13.8	6.3	0.2	Tr	100	17.2	0.20	65	4.5	7.7	–

(continued)

10.3 General Characteristics of Urban Soils

Table 10.6 (continued)

Horizon	Texture	Sand %	Silt %	Clay %	Exchangeable Bases Ca cmol kg^{-1}	Mg	K	Na	Base saturation %	Org. C %	Total N %	C/N	Bray # 1 P mg kg^{-1}	pH	Bulk density g cm^{-3}
3^Bwu	LS	84	11	5	5.6	1.5	0.1	–	100	1.2	0.01	118	8.3	7.8	–
^Bw	LS	86	10	4	2.1	0.6	0.1	–	100	0.2	0	0	28.5	7.7	–
3Ab	SL	78	14	8	16.0	3.2	Tr	Tr	100	3.1	0.3	12	7.0	7.6	–
3Bgb	LS	80	14	6	5.7	1.3	–	Tr	100	0.6	0.05	13	3.9	7.8	–
3Bwb	SL	75	15	9	5.8	1.5	Tr	Tr	100	0.2	0.02	11	0.7	8.0	–
3BCb	S	90	6	4	3.4	0.6	–	–	100	0.2	0.02	12	2.6	8.1	–
3C	S	96	3	1	28.3	0.8	–	0.3	100	1.1	0	0	0.8	8.6	–
Laguardia series (Anthroportic Udorthent)—Kings County, New York															
^Au	SL	66.5	24.7	8.8	19.9	1.3	–	0.2	100	3.91	0.09	38	218.2[a]	7.5	1.43
^BCu	SL	65.0	25.7	9.3	27.7	1.0	–	0.1	100	3.95	0.04	100	81.8[a]	7.9	1.50
^Cu	SL	71.6	20.4	8.0	24.0	0.9	–	0.1	100	5.53	0.07	78	34.9[a]	8.1	1.54
Rikers series (Anthroportic Udorthent)—Richmond County, New York															
^Au	S	78.9	14.7	6.4	9.3[a]	1.3	–	0.3	100	4.98	0.31	16	110.6	5.5	1.49
^Cu1	S	79.6	16.9	3.5	16.2[a]	0.6	Tr	0.2	100	15.11	0.25	60	28.5	7.0	1.06
^Cu2	S	82.1	11.3	6.6	8.1[a]	0.3	–	0.1	100	4.11	0.16	26	56.0	6.6	1.54
^Cu3	SL	78.3	8.8	12.9	4.2[a]	0.2	–	0.1	100	1.13	0.04	28	56.0	6.8	–

[a]Mehlich #1 method

Compared with the natural soil, levels of organic C, total N, pH, base saturation, and exchangeable Ca, Mg and K are significantly higher in the urban soils. Levels of Bray #1-extractable P were also higher ($\sim$11–48 mg kg^{-1}) than that of the typical unfertilized natural soil in Michigan ($\leq$ 12 mg kg^{-1}). Two of the Michigan soils had P-levels within the optimum range (16–22 mg kg^{-1}) for many crops, although the Bray #1 test is not considered to be appropriate for calcareous soils. In contrast to the natural soil, which shows a regular decrease in organic C with increasing depth, the anthropogenic soils show the irregular distribution with depth that has been widely recognized in urban soils. Given that natural soils in woodland areas generally contain 2–3% organic C, whereas floodplain and grassland soils may contain as much as 5–6% organic C, the data also suggest that levels of organic C have rebounded to levels similar to those of natural soils in 70 years or less. The anthropogenic soils also have re-established C/N ratios similar to that of the natural soil, although the organic C levels measured in the anthropogenic soils (using the combustion method) were likely affected by the presence of carbonaceous artifacts.

In contrast to these data, the surface layer of recently deposited HTM typically contains much less than 0.6% organic C (unless topsoil was replaced by a demolition contractor). Total levels of N, P and S are correspondingly low in young urban soils, although P may be elevated locally by the presence of animal bones or midden materials. Cation exchange capacity is typically low, as a result of low organic matter content, but this may be offset to some extent by a greater component of pH-dependent CEC at higher pH. The pH of HTM-type anthrosoils is typically elevated, mainly due to the presence of calcareous artifacts and microartifacts such as concrete, and mortar, as well as certain types of brick, coal combustion products, glass and steel-making slag (Meuser 2010; Howard and Orlicki 2015). Levels of exchangeable Ca and K may be higher in urban anthrosoils than in associated natural soils as a result of calcareous artifacts, and the incorporation of unweathered sediment into the fill material. Deicing salts and gypsum (derived from plaster and drywall) can generate high levels of soluble salts in urban soils locally, but these levels may rapidly dissipate as a result of leaching. The impact of deicing salt could hypothetically be indicated by anomalously high levels of exchangeable Na, but the excess of Ca may sometimes mask this effect. The chemical properties of urban anthrosoils are often heterogeneous as a consequence of their highly variable soil profile and geospatial characteristics.

The type and amount of organic matter generally controls soil biological activity, and greatly affects the soil biota comprising the food web. Microbes and earthworms and other biota feed on soil organic matter, hence recently deposited HTM lacking an ^A horizon is characterized by greatly diminished biological activity. Microbial C, N and P all decline with increasing human disturbance (Chen et al. 2001; Yang et al. 2001). Urban soils are low in microbial biomass carbon (MBC) and respiration, which limits C, N and P cycling (Vauramo and Setala 2011). MBC normally comprises 1–5% of total organic C (Wang et al. 2011). In China, cropland typically has an MBC of 100–600 mg C kg^{-1} and pasture may contain up to 1500 mg C kg^{-1} (Wang et al. 2011; Chen et al. 2013). This amount is

10.3 General Characteristics of Urban Soils

reduced by 50% or more in urban soils (Scharenbroch et al. 2005; Chen et al. 2013). Microbial C and respiration have been found to increase immediately after human disturbance of the soil, and then decrease. This is interpreted to indicate diminished organic matter accumulation, and losses in both labile and stable (recalcitrant) forms of total C. The decrease in microbial biomass and activity is attributed to a lack of organic matter and limited nutrient supply (Zhao et al. 2013). The decrease in microbial respiration, biomass C, N and P, and particulate and other forms of total C, are attributed to diminished organic matter content, but urban soils have been found to rebound to quasi-natural conditions in less than 100 years. The presence of earthworms is favored by the basic pH of many urban anthrosoils (Burghardt 1994). Scharenbroch et al. (2005) found that biological conditions similar to those of natural soils had been achieved in urban soils within ~60 years following human disturbance.

Compared with natural soils, the geophysical properties of urban soils can be significantly altered by the presence of artifacts and microartifacts (Table 10.7). Howard and Orlicki (2015) found that the electrical conductivity (EC) of natural soil A horizons was <110 $\mu S\ cm^{-1}$, whereas that of anthropogenic surface horizons averaged 335–468 $\mu S\ cm^{-1}$, and were therefore about three to four times the average background level. Calculated resistivities suggested that natural soils were only weakly corrosive to metals, but anthropogenic soils were moderately to strongly corrosive to metals. Calcareous and ferruginous (corroded nails) building material wastes probably caused elevated ECs at demolition sites, whereas fly ash-impacted and industrial soils were probably affected more by the ferruginous components of coal-related and steel-making wastes. Similarly, natural A horizons had a magnetic susceptibility (MS) $<83 \times 10^{-8}\ m^3\ kg^{-1}$, whereas MSs at residential and fly ash-impacted sites averaged 113 and $179 \times 10^{-8}\ m^3\ kg^{-1}$, respectively, and were thus two to three times the background level. The average MS of industrial site soils was 1493, about 20 times the background value. Magnetic microspheres were found to be abundant in fly ash-impacted and industrial site soils. Hence, these and other ferruginous wastes likely contributed to the elevated values of MS (Table 10.7). Anthropogenic soils in many parts of the world have been found to have elevated values of EC and MS, especially those associated with metallurgical wastes (Howard and Orlicki 2015). Fly ash-impacted soils with elevated values of MS have been recognized previously at distances of 10 km or more from coal-fired power plants. The elevated MS associated with fly ash-impacted soils was attributed to the presence of highly magnetic ferruginous microspheres. Magnetite microspheres are known to be produced as a by-product of coal combustion. Artifacts with high specific gravities, such as hematite and magnetite, may contribute to the high bulk densities of urban soils. This includes metalliferous slag, corroded iron nails, coal cinders and glass slag (produced by iron smelting). The specific gravity of artifacts may be generally categorized as ferruginous > calcareous $\geq$ siliceous > carbonaceous.

Soil surface layer temperatures generally vary more or less according to the temperature of the air, but these layers are generally warmer than the air. Heat

Table 10.7 Chemical and geophysical characteristics of natural versus anthropogenic topsoils in the metropolitan Detroit, Michigan area as a function in land use type (Howard and Orlicki 2015)

Site	pH X	pH S	pH CV (%)	Electrical conductivity (μS cm^{-1}) X	S	CV (%)	Electrical resistivity (ρ) (Ωm) ρ	Corrosivity index	Magnetic susceptibility (10^{-8} m^3 kg^{-1}) X	S	CV (%)
Native soils (parkland or farmland)											
1	6.96	0.08	1.2	109.5	9.7	8.9	91.3	Weak	83.2	5.3	6.4
2	7.24	0.06	0.9	189.5	30.1	15.9	52.7	Weak	17.2	0.4	2.2
3	4.59	0.02	0.5	75.30	0.8	1.0	132.8	Very weak	12.7	0.1	0.1
4	6.48	0.35	5.5	92.70	1.1	1.2	107.9	Very weak	57.2	1.2	2.0
Anthropogenic soils (residential demolition)											
5	8.20	0.35	4.3	240.2	129.8	54.0	41.6	Moderate	68.8	0.6	0.8
6	7.96	0.12	1.5	304.3	92.2	30.3	32.9	Moderate	57.5	13.5	23.5
7	7.60	0.11	1.4	359.0	68.4	19.1	27.9	Moderate	167.0	2.1	1.3
8	7.68	0.01	0.9	221.6	35.5	16.0	45.1	Moderate	115.7	4.8	4.2
9	7.52	0.07	0.9	614.3	134.8	21.9	16.3	Strong	131.5	8.0	6.1
10	7.38	0.19	2.6	876.7	128.9	14.7	11.4	Strong	249.7	2.6	1.0
11	7.65	0.21	2.77	259.2	46.5	17.9	38.6	Moderate	174.6	3.6	2.0
12	7.80	0.03	0.36	303.1	24.2	8.0	33.0	Moderate	68.3	22.5	32.9
13	7.51	0.04	0.47	1138.0	834.9	73.3	8.8	Strong	30.1	16.0	53.1
14	7.94	0.02	0.27	358.9	27.1	7.6	27.9	Moderate	64.9	9.9	15.3
Anthropogenic soils (industrial)											
15	7.86	0.14	1.8	398.5	12.0	3.1	25.1	Moderate	70.2	2.3	3.3
16	7.74	0.01	0.2	354.5	34.7	9.8	28.2	Moderate	1857.0	8.0	0.4
17	8.75	0.01	0.1	168.8	0.4	0.3	59.2	Weak	1183.0	73.8	6.2
18	7.83	0.06	0.7	433.7	12.7	2.9	23.1	Moderate	2436.0	9.5	0.4
19	7.77	0.02	0.3	320.9	8.7	2.7	31.2	Moderate	1918.0	68.1	3.6
Anthropogenic soils (fly ash-impacted)											
20	7.13	0.07	1.0	539.9	10.1	1.9	18.5	Strong	94.8	4.9	5.2
21	7.83	0.03	0.4	296.0	1.4	0.5	33.8	Moderate	90.5	2.4	2.6
22	7.80	0.09	1.2	483.8	45.5	9.4	20.7	Moderate	296.5	4.6	1.5
23	7.98	0.01	0.1	208.9	22.3	10.7	47.9	Moderate	302.5	1.7	0.6
24	7.72	0.01	0.2	236.2	8.2	3.5	42.3	Moderate	111.0	0.1	0.0
25	7.81	0.06	0.7	159.1	24.8	15.6	62.8	Weak	158.4	6.0	3.7

X mean; *S* standard deviation; *CV* coefficient of variation

passes through water many times faster than air, and heat moves through mineral particles even faster than water, so when particle-particle contact increases through compaction, heat-transfer rates are increased. Bare soil warms and cools faster and is more affected by frost penetration. Thus, human disturbance, especially devegetation causes urban topsoils to be generally warmer than those in rural areas. Also, the air temperature is generally greater due to the urban heat island effect. The mean annual air temperature of a city with 1 million people or more can be as high as 2–5 °F (1–3 °C) warmer than surrounding rural areas, and the difference can be even greater at night. Thus, urban topsoils are significantly warmer than non-urban soils. The presence of sealed urban soils reduces the cooling effects of evapotranspiration, and soil may be surrounded by large heat-absorbing and reradiating structures, further increasing air and soil temperature. Soil temperature affects the moisture content, thus urban soils may be drier than their natural counterparts. It also controls the growth microenvironment of roots and soil macrofauna, and affects inorganic chemical processes. Higher temperatures in urban soils imply higher inorganic and organic reaction rates, and therefore increased rates of organic matter decomposition and intensified weathering. This partly explains reports of accelerated weathering of artifacts, and C and other nutrient losses. Excessive compaction and soil sealing can significantly impact the hydrological characteristics of urban areas, and thus affect the soil moisture regime. Although compaction and sealing generally act to cause a lowering of the urban water table, the opposite effect can also be observed as a result of reduced evapotranspiration rates (Burghardt 1994).

In urban areas, soil crusts may be characteristic of bare ground surfaces and typically form where vegetation has been destroyed or removed during the course of construction, demolition, and other human activities. They can be classified into physical, biological and chemical types. Physical crusts form as a result of raindrop splash, freeze-thaw action, piping, and water repellency. A physical crust may form when falling drops of rain break apart aggregates, and then fines wash into and fill voids amongst the coarser fraction. The surface becomes covered with a thin layer of fine structureless material called a seal. Even a minor amount of clay can cause the seal to harden into a crust upon drying. Other physical crusts form in silt-rich substrates by piping as water puddles or ponds on the surface either when the rate of rainfall greatly exceeds infiltration, or during freeze-thaw cycling. Piping occurs when infiltrating water traps air between the larger voids associated with the sand and gravel fraction, and escaping air bubbles carry silt and/or clay particles upward into the ponded water column where they settle out of suspension and accumulate on the soil surface. Water repellency results from waxy organic compounds (Doerr et al. 2000) which coat soil particles to form a hydrophobic, crust-like layer as much as 3 cm thick which lies just beneath the soil surface. The hydrophobic substances are derived from certain types of plant materials forming a litter on the ground surface, or they may be derived from organic substances excreted by fungi. The waxy substances sometimes penetrate into the soil as a result of fire. Coarse textured sandy soils containing <5% clay are most susceptible to becoming water repellent. Physical crusts generally form on, or within a few centimeters of, the surface and may be 5 cm or more in thickness. They can greatly restrict infiltration (leading to increased surface runoff and erosion), aeration, evaporation, and seedling emergence. Biological and chemical crusts are most common in regions of semi-arid to arid climate (Eldridge and Greene 1994). Biological crusts (also known as microbiotic crusts) are comprised of a complex community of lichen, cyanobacteria, algae and moss which grow together and bind the soil with a matrix of organic fibers and exudates. Chemical or salt crusts are formed on saline soils by the chemical evaporation of salts on the soil surface.

10.4 Urban Soils on Residential Land

Anthropogenic soils on modern residential land are formed in HTM typically containing a diverse artifact assemblage reflecting human habitation. Artifacts, in the form of waste building materials related to construction or demolition, are ubiquitous in residential urban soils. Artifacts may also include coal-related wastes as a legacy from the era of domestic coal use, as well as all manner of household items. The kind and amount of demolition debris depends on the age and type of building construction, as well as variations in the methods used by different demolition contractors, and changes in municipal regulations governing them. Brick and mortar construction generally produces the greatest volume of demolition debris. The kinds of artifacts vary according to the evolution in building materials over time. For example, hand-forged iron nails and other objects are common at demolition sites where buildings from the 19th century were demolished. Mortar was produced from natural hydraulic cement during the 19th century, in contrast to the Portland cement commonly used during the 20th century. Such compositional differences affect urban soil mineralogy and associated chemical properties, and artifacts may weather at different rates depending on the time period during which they were produced (See Chap. 5).

In modern cities where urban redevelopment is occurring, it is common practice to demolish old buildings. Most of the waste remaining after demolition is hauled off to be recycled, scrapped or landfilled. Earthmoving equipment is then used to grade and level the site with earthy fill, which is often imported from a nearby location offsite, along with any remaining demolition debris. The HTM created by this process is typically a mixture of soil and/or sediment containing variable quantities of artifacts. Howard and Shuster (2015) carried out a study in Detroit, Michigan to determine if soils could be mapped on a vacant lot <0.1 ha in size created by demolition of a single family home dating from ∼1925. The original footprint of the house was evident from ground subsidence (Fig. 10.8a). Data collected via a grid of 90 hand-auger borings showed that a mappable, albeit complex, pattern of soils was present (Fig. 10.8b). There were also well defined patterns in the spatial distribution of artifacts with respect to the former location of the house. For example, wood artifacts were concentrated in the vicinity of the basement resulting in ground subsidence. The excavational history could be inferred from cross sections constructed across the site (Fig. 10.8c). Overall, soils became progressively less disturbed, grading from Urbic Technosols at the former house site, to Hortic-like Anthrosols in the former house yard, and then into natural

10.4 Urban Soils on Residential Land

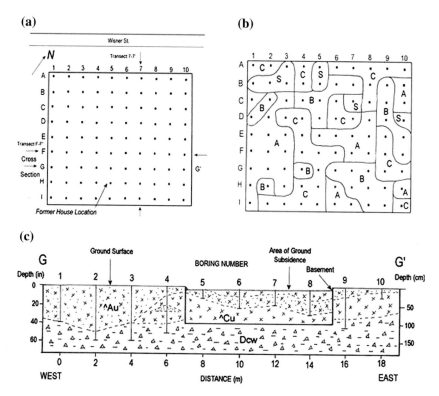

Fig. 10.8 Experimental mapping of demolition site. a Sampling grid in relation to former house location indicated by ground subsidence. b Soil map showing configuration of four distinct soil types (A, B, C, S). c Cross-section though site showing anthrostratigraphy in relation to area of ground subsidence and basement marking former location of house. *Dcw* clayey diamicton parent material (Howard and Shuster 2015)

soils, with increasing distance from the demolished home. In contrast to this demolition site dating from ∼1997, which was characterized by deposits of fill ≥1 m thick, mapping of another demolition site in Detroit dating from 1918 showed the presence of only a very thin cover of earthy fill <30 cm thick. This was attributed to the fact that earthmoving was done primarily by human and animal labor, or perhaps using primitive steam-powered tractors (Howard et al. 2015). Further mapping in Detroit of numerous undemolished and demolished first-generation homes, showed a similar pattern of anthropogenic soils. Thus, sites pre-dating the advent of diesel-powered earthmoving equipment during the 1930s showed relatively little construction-related disturbance of the natural soil, and had relatively thin covers of fill, compared with more modern sites.

Table 10.8 compares the properties of chronosequences of natural and demolition site soils in the Detroit, Michigan area. The sandy natural Holocene floodplain topsoil was highly calcareous, organic-rich, and had high levels of exchangeable

Table 10.8 Physical and chemical properties of natural versus anthropogenic soil chronosequences near Detroit, Michigan, USA

Horizon	Texture	Sand %	Silt %	Clay %	Ca	Mg	K	Na	CEC	Base sat. %	Org. C %	Total N %	C/N	pH	Carbonate %
					cmol kg^{-1}										
Aeric Fluvaquent (floodplain, ~5000 year old)															
A1	Sil	41	53	6	23.2	1.2	0.03	Bdl	31.0	79	4.2	0.18	22	7.8	9.8
A2	Sl	52	42	6	19.9	0.9	0.02	Bdl	28.0	74	3.6	0.12	31	7.9	14.4
C	S	91	8	1	11.8	0.5	0.01	Bdl	16.0	74	1.3	0.04	35	7.9	2.4
Typic Dystrochrept (beach ridge, ~12,400 year old)															
Ap	S	88	7	5	4.9	1.0	0.09	0.03	06.0	100	1.4	0.10	14	5.7	0
AB	S	88	8	4	2.6	0.6	0.11	0.02	04.5	73	0.5	0.02	18	6.2	0
Bw	S	91	4	5	1.8	0.4	0.09	0.02	02.4	100	0.3	0.02	17	6.6	0
BC	S	95	2	3	0.9	0.2	0.20	0.02	01.2	100	0.2	0.01	19	6.5	0
C	S	97	2	1	0.5	0.1	0.05	0.01	01.7	39	0.1	Bdl	19	6.5	0
Typic Hapludalf (beach ridge, ~13,000 year old)															
A	S	93	7	0	12.8	1.6	0.06	Bdl	15.0	100	2.8	0.24	12	6.5	0.9
Bt	Ls	87	4	9	7.5	0.6	Bdl	Bdl	08.0	100	0.7	0.04	20	7.1	2.9
BC	Ls	86	8	6	3.6	0.5	0.02	Bdl	04.1	100	0.2	0.02	10	7.1	0.1
C	S	96	4	0	13.4	0.3	0.03	Bdl	14.0	100	0.2	0.01	131	8.1	10.4
Anthroportic Udorthent (~3 year old)															
^Cu	Scl	66	14	20	17.1	0.2	1.1	0.04	18.5	100	4.0	0.10	24	8.0	2.2
2^Cdu	Sl	69	18	13	21.2	0.2	0.9	0.11	22.9	98	5.8	0.10	56	8.5	10.5
3^Cdku	–	–	–	–	–	–	–	–	–	–	–	–	–	8.2	–
Anthrodensic Udorthent (~24 year old)															
^Au	Sl	53	34	13	18.2	0.4	2.0	0.02	20.9	99	2.7	0.14	20	7.9	2.3
^Cdu	Sl	54	27	19	16.2	0.2	2.6	0.03	19.5	98	1.7	0.06	31	8.1	1.8

(continued)

10.4 Urban Soils on Residential Land

Table 10.8 (continued)

Horizon	Texture	Sand %	Silt %	Clay %	Ca	Mg	K	Na	CEC	Base sat. %	Org. C %	Total N %	C/N	pH	Carbonate %
					cmol kg^{-1}										
Anthropic Udorthent (~39 year old)
^Au	L	53	32	15	19.3	0.4	2.1	0.04	21.9	100	4.2	0.03	14	7.7	1.9
^Cku	Cl	34	36	30	22.3	0.2	2.3	0.15	25.3	98	2.1	0.07	29	8.0	7.8
2Ab	L	38	37	25	15.7	0.2	3.1	0.31	19.3	100	1.2	0.13	9	8.1	1.3
2C	L	34	31	35	12.4	0.2	3.0	0.39	16.0	100	0.5	0.06	8	7.8	3.8
Anthroportic Udorthent (~68 year old)
^Au	Sl	76	15	9	14.8	0.3	1.1	0.06	21.9	100	3.2	0.25	8	7.7	1.7
^Cu	Cl	37	35	28	20.7	0.2	1.8	0.08	25.3	98	2.1	0.08	8	8.0	6.4
2Ab	Sl	71	18	11	15.6	0.3	1.2	0.06	19.3	100	3.1	0.25	9	8.1	1.7
2Cg	Scl	51	26	23	7.3	0.3	1.4	0.07	16.0	100	0.3	0.05	8	7.8	0.9

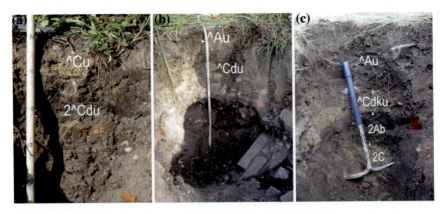

Fig. 10.9 Chronosequence of demolition site soils (Detroit, Michigan USA). **a** 3 year-old soil with ^Cu profile. **b** 24 year-old soil with ^Au-^Cu profile. **c** 68 year-old soil with anthropic epipedon. Scales marked in 10 cm intervals. Photos by J. Howard

bases. The sandy soils developed in late Pleistocene lacustrine beach ridges showed increasing morphological development, grading from Inceptisols to Alfisols with increasing soil age (Howard et al. 2012). They were leached of carbonate to a depth of about 1.2 m, and had lower levels of organic C, pH and exchangeable bases than the floodplain soil, but had a base saturation >50%. In contrast, the anthropogenic soils had clay contents that were four or five times higher, which can be attributed to mixing of sandy and clayey fill materials during the demolition operation. They had higher levels of pH, organic C, and exchangeable Ca, K and Na, than the beach ridge soils. They were characterized by higher carbonate contents due to the presence of abundant calcareous artifacts. Some of the urban soils had organic C contents equal to or greater than that of the organic-rich floodplain soil (Howard et al. 2013). The demolition site soils initially showed no signs of melanization, but an ^A horizon had developed within ∼24 years (Fig. 10.9). ^A horizons systematically increased in thickness with increasing soil age (Fig. 10.10a). The soils exhibited a systematic increase in total nitrogen content, whereas C/N ratios decreased, with increasing soil age (Fig. 10.10b, c). There was also a progressive decrease in pH with increasing soil age (Fig. 10.10d). Organic C showed no trend with time, which was attributed to the presence of coal-related wastes and the use of the combustion method (Howard et al. 2013). However, another chronosequence study in the same area using the Walkley-Black method suggested that soil organic matter content increased from 2.1% in a 16 year-old demolition site soil, to 6.1% after 120 years of development (Howard and Olszewska 2011). Accelerated weathering of mortar appeared to be taking place within about 15 years, attributed to the effects of deicing salts and other human activities, and weathering of artifacts had resulted in measureable increases in carbonate and Fe-oxide content after 20–40 years of soil development. The anthropogenic soils had higher levels of

10.4 Urban Soils on Residential Land

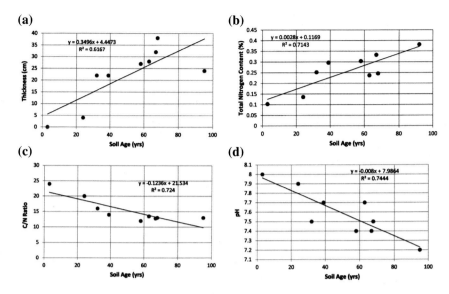

Fig. 10.10 Chronofunctions of various demolition site soil A or surface horizon characteristics. Modified from Howard et al. (2013)

plant-available K because K-fixation was occurring in the natural soils as a result of weathering of muscovite and vermiculite (Howard et al. 2012). Overall, the results suggested that anthropogenic soils could be more fertile than natural soils, and the presence of artifacts could in some cases have beneficial effects. The urban soils studied were resilient, having re-established a C/N ratio typical of natural soils (i.e. C/N = 12–13) within 60–100 years of development. Similar resilience in urban soils has been recognized elsewhere in the United States (Scharenbroch et al. 2005).

Prokofeva et al. (2001) studied Russian soils sealed under asphalt, concrete, and other types of road covering dating to the 16th, 18th and 19th centuries. They found that such "ekranozems" were characterized by extensive evidence of dissolution and illuviation of pedogenic carbonate. They also documented pedogenic gypsum in the sealed soils, in some cases with calcite pseudomorphs over gypsum. Unsealed urban soils (urbanozems) in the same area showed much less evidence of carbonate redistribution. Puskas and Farsang (2009) also found highly calcareous soil beneath technic hard material, but no mention was made of translocation of carbonate. Shaw and Reeve (2008) studied urban soils adjacent to a car park that had been unpaved for about 20 years. Nutrient levels, pH and soluble salt contents of surface soils all increased toward the car park, indicating an impact of human activities on natural soils. It is well established that deicing salts can affect the chemical properties of urban soils, and the effects can extend for tens of meters beyond the highway (Cunningham et al. 2008).

10.5 Urban Soils on Industrial Land

Industrial land, as defined here, includes sites of heavy industry, land modified by mining and dredging operations, land utilized for solid waste disposal (landfills), and land dedicated to transportation-related activities (highways, railroads, airports). Mine soils were discussed in Chap. 9, and landfill and dredged land soils were described previously by Meuser (2010). Hence, the discussion here will focus on soils affected by heavy industry, and on the impacts of coal-related wastes.

Heavy industries are generally those which sell their products to other industries, rather than to end users and consumers, i.e. they make products that are used to make other products. Heavy industry typically involves the use of considerable heavy machinery, equipment, and physical plant to produce large amounts of standardized products, often utilizing mass production or flow production on assembly lines. Examples of heavy industry include oil production and refining, shipbuilding, steel-making, chemical, automobile and machinery manufacturing, paper mills, coal-fired power plants, foundries, coking operations and manufactured gas plants, and other commercial operations. An industrial site is usually relatively small, often less than a few hectares in size, but large sites may cover 250–800 ha. In the United States, the typical site dates from the late 19th or early 20th century, when many different industries were often clustered together in a heavily industrialized terrain. The sites were invariably located near a waterway, and characterized by stockpiles of raw materials, such as iron ore, coal or coke, limestone, sand, chromite or other metallic ores; waste piles comprised of coal cinders, steel-making slag, etc.; and settling ponds containing waste sludge or tailings. Today, a former industrial site is often characterized by an abandoned, derelict building, or vacant land produced by building demolition. Rubble, wastes and contaminants are distributed indiscriminantly about the site. Thick deposits of HTM, perhaps 6–10 m or more in thickness, are present, and typically contain coal-related wastes (cinders, ash, tar, etc.), steel-making slag, waste building materials produced by building demolition, and various hazardous wastes that were spilled or dumped on the site. The sites often have the remains of old foundations, buried drains, pipelines, corroded underground storage tanks, and layers of technic hard material covered in earthy fill. In contrast to HTM in residential areas, where deposition is primarily a one-time occurrence, HTM deposition at industrial sites may have occurred in many episodes over time. For example, ongoing production of manufacturing wastes could result in daily or weekly depositional events, producing stratified deposits of HTM. If long gaps in time occurred between depositional events, previous weathering events may become preserved beneath subsequently deposited layers, similar to the situation commonly observed in natural floodplain soils (Huot et al. 2015).

Industrial site soils are typically characterized by ^Au-^Cu profiles (Huot et al. 2015). ^A horizon development and organic C content can be highly variable, and patchy, if the HTM contains substances which are phytotoxic. Huot et al. (2013) reported that blast furnace sludges had developed an ^A-^C profile in 24 years, and

Scalenghe and Ferraris (2009) noted the development of an ^A horizon in retaining wall fill, along with minor leaching and pH drop, after only 4 years of pedogenesis. Sere et al. (2010) also reported that soil profiles formed in paper mill sludge had developed structure and showed signs of minor leaching of soluble compounds within 3 years. Some industrial site soils have also been reported to have cambic-like ^B horizons based on the development of weak structure and some leaching of soluble salts and carbonate (Huot et al. 2015). Scalenghe and Ferraris (2009) observed the development of a cambic or cambic-like ^B horizon in retaining wall fill after 40 years of pedogenesis. Thus, industrial site soils have been reported to show signs of accelerated weathering similar to those observed in mine soils and residential urban soils.

The industrial soil profile shown in Fig. 10.11a was located adjacent to a foundry in a formerly heavily industrialized part of Detroit City. The upper part of the profile contained an artifact assemblage characterized by wood, brick, mortar, lime and asphaltic concrete, cinder block, coal, tar, coal cinders and metalliferous slag with little fine earth material. Corrosion of ferruginous artifacts had created an abundance of bright yellow ferrihydrite. The underlying dark-colored layer contained artifacts typical of foundry wastes including, coal, coked coal, coal cinders, metalliferous slag and unidentified ferruginous materials, in a matrix of soil particles weakly cemented by coal tar. Microscopic analysis revealed the presence of microspheres and microagglomerate typical of coal ash (Howard and Orlicki 2016). Figure 10.11b shows a typical profile of the Laguardia soil, which is a common anthrosoil in the New York City area. It was developed in a deposit of HTM >2 m thick containing abundant artifacts (brick, mortar, asphalt, etc.). The Mosholu soil (Fig. 10.11c) was mapped in association with landfill sites in New York City. Industrial soils are typically very dark colored as a result of black carbon microparticles, and often have a skeletal texture containing very little fine earth fraction (Burghardt 1994).

Industrial soils associated with foundries and steel-making operations often contain slag, a rock-like waste material produced by the smelting of metalliferous ores (Fig. 10.12). There are three types of iron industry slag (Proctor et al. 2000): (1) blast furnace (BF) slag, (2) basic-oxygen furnace (BOF) slag, and (3) electric arc furnace (EAF) slag. BF slag is produced by the smelting of iron ore to produce cast-iron, whereas BOF and EAF slags are generated during steel-making. All three types are comprised of fluxing agents (Ca and Mg lime) and the molten impurities of iron ore. They are similar in chemical composition, but steel-making slags are generally higher in Fe and Mn. Smelting is a process in which heat and a chemical reducing agent are used to selectively extract a metal by melting and driving off other elements as gasses or slag. The reducing agent is commonly carbon in the form of coke. The carbon compounds remove oxygen from the ore, leaving elemental metal behind. Most ores contain silicate and other mineral impurities, hence it is usually necessary to use a flux to remove the accompanying rock gangue as slag. During a typical steel-making operation, Fe-ore, coke and flux (limestone or dolostone) are heated to ~ 2000 °C in a refractory brick-lined blast furnace.

Fig. 10.11 Profiles of anthropogenic soils (Technosols) in industrial settings. **a** Demolition site soil (Spolic Technosol) at former foundry (Detroit, Michigan). Dark layer comprised of slag and coal cinders weakly cemented by coal tar. **b** Deposit of human-transported material >2 m thick comprising the Laguardia series (Spolic Technosol; Kings County, New York). **c** Landfill soil (Garbic Technosol) possibly correlated with the Laguardia series (Bronx, New York). **d** Coal-combustion products comprising the Mosholu series (Bronx, New York). Photos by J. Howard and USDA-NRCS. Scale in cm

10.5 Urban Soils on Industrial Land

Fig. 10.12 Steel-making slag near Baden, Pennsylvania, USA. Fragments range from metalliferous, with basalt-like appearance, to a glassy and obsidian-like. Photos by Russell Losco

Quicklime is formed by calcination of calcite which reacts with silica to produce slag as follows:

$$CaCO_3 \rightarrow CaO + CO_2 \tag{10.1}$$

$$CaO + SiO_2 \rightarrow CaSiO_3 \text{ (wollastonite)} \tag{10.2}$$

The liquid slag floats on the molten iron and is separated by decantation. The flux is added to remove impurities (e.g. sulfur and phosphorous) and lower the viscosity of the slag. Ordinary blast furnace slag is typically a mixture of metal oxides and silica in the form of glass. However, the glass phase is much less common in slag produced during the more advanced stages of refining. Common mineral components include magnetite, hematite, wustite, merwinite, melilite, forsterite, monticellite, and wollastonite (Muralha et al. 2011; Yildirim and Prezzi 2011). Slag is characteristically high in lime, which reacts readily with water to produce portlandite and then undergoes carbonation:

$$CaO + H_2O \rightarrow Ca(OH)_2 \tag{10.3}$$

$$Ca(OH)_2 + CO_2 \rightarrow CaCO_3 + H_2O \tag{10.4}$$

These reactions give calcareous steel slag its cementitious properties (Tasakiridis et al. 2008). Slow cooling of slag produces an unreactive Ca-Al-Mg silicate, but rapid cooling below 800 °C produces a granular material with the properties of sand. This material is often ground into a fine powder (ground-granulated blast furnace slag) which is combined with aggregate to produce concrete (cinder) block and other building materials. Slag is known to sorb heavy metals (Kim et al. 2008; Oh et al. 2012), and calcareous slag produced by the Bessemer or Linz-Donowitz processes may have agricultural value because of its high lime and phosphorous content (Kim et al. 2012). However, its exceptionally high pH (8–10) makes revegetation of steel slag difficult (Cremeens et al. 2012). In the past, steel slag was marketed locally as good fill for construction sites.

Highly calcareous industrial soils can be generated as a byproduct of chemical manufacturing (e.g. the Solvay process), manufactured gas operations, flue-gas desulfurization, and fluidized bed combustion. The Solvay process was invented by Ernest Solvay in the 1860s for the production of soda ash, which was originally produced from the ash of certain plants. The process utilizes common ingredients, such as salt brine and limestone, and produces sodium carbonate as follows:

$$2NaCl + CaCO_3 \rightarrow Na_2CO_3 + CaCl_2 \tag{10.5}$$

Sodium carbonate is widely used for the manufacturing of glass, paper, soaps and detergents. For example, bottle and window glass are comprised of Na-lime glass produced by melting Na_2CO_3, $CaCO_3$ and SiO_2. The Solvay process and purification of salt brine generates a highly calcareous, fine, powdery waste, known

10.5 Urban Soils on Industrial Land

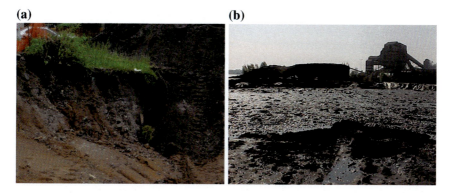

Fig. 10.13 Contaminated urban soils. **a** Contaminated soil (*black*) at former industrial site (Connecticut). Photo by USDA-NRCS. **b** Soil at former coking plant (Detroit, Michigan) polluted with coal tar sludge. Photo by Michigan Department of Environmental Quality

locally as "distiller blowoff," comprised of $CaCl_2$ and $CaCO_3$. The Solvay process is not used much today because extensive deposits of Na carbonate, in the form of trona, are now known in Wyoming and California. Industrial anthrosoils can be impacted by distiller blowoff, or by highly acidic brine waste from soda ash production, as a result of accidental or intentional dumping. Very alkaline (pH 11–12) muddy sludges are also known to be produced by the refining of aluminum from bauxite (Santini et al. 2013). Highly toxic organic wastes generated by the chemical manufacturing industry, as was the case at the infamous Love Canal site, are also common in industrial anthrosoils (Fig. 10.13a).

10.6 Urban Soils Impacted by Coal-Related Wastes

Coal fueled the Industrial Revolution, and is still used extensively in steam engines and turbines, and for most industrial purposes. Urban soils have been widely affected by coal-related wastes in the form of coal combustion products (Fig. 10.11d), unspent coal, carbonaceous shale, and coal-tar. Fly ash is composed of the sand-sized and finer mineral matter that remains after coal burning, along with carbonaceous products of incomplete combustion. Urban soils are often impacted by fly ash derived from coal combustion associated with iron smelting foundries, steel-making mills and other heavy industry, coal-fired power plants, manufactured gas plants, and domestic coal use. In the United States, the widespread use of coal as an energy source began about 1850, but in Europe coal had been used extensively since the initiation of the Industrial Revolution at the beginning of the 18th century. The air pollution and adverse health effects created by coal combustion were so extensive that they affected the structure of British cities, as discussed previously in Sect. 10.1, eventually culminating in the well known London smog event of 1952. Extensive areas of fly ash-contaminated soils are known in association with heavily industrialized regions in Germany and Poland. The adverse effects of coal combustion have also grown enormously in Japan and China since the end of WWII, as seen in the recent Shanghai smog event of 2013. Fly ash contamination can be in the form of waste materials left onsite following demolition, or in settling ponds, but often results from airborne deposition. Urban soils are mainly affected by airborne deposition of fly ash at distances <10 km from a point source, but can be affected at distances of 10–30 km or more (Schmidt et al. 2000; Koschke et al. 2011). Airborne deposition of fly ash mainly affects the surface layer (0–10 cm depth), but silt-sized and finer fly ash particles are prone to illuviation and may also be found deeper in the soil profile.

Fly ash can have significant effects the physical, chemical, mineralogical, geophysical and biological characteristics of urban anthrosoils. Even at low levels of contamination, fly ash accumulation can result in increased porosity and water-holding capacity. Increased hydraulic conductivity and decreased water repellency, may lead to reduced plant-available water. Fly ash promotes macroaggregation, and at high levels of contamination, fly ash can cause a marked reduction in bulk density (Hartmann et al. 2010; Yunusa et al. 2011). Fly ash can cause a significant increase in pH, exchangeable Ca, base saturation, and CEC (Schmidt et al. 2000; Koschke et al. 2011). Weathering of fly ash can cause further impacts on urban soil chemistry. For example, fly ash has been shown to release significant levels of Ca and SO_4 after only about 20 years of weathering (Zikeli et al. 2002). The microspheres and cenospheres which are characteristic of fly ash are comprised of black carbon, aluminosilicate glass and magnetite. Large quantities of glass in fly ash-impacted soils can impart andic-like properties, similar to those associated with volcanic soils. However, fly ash-impacted soils differ from their volcanic counterparts in having much higher levels of gypsum, carbonate, and total organic carbon. Total organic carbon levels as high as 52% having been measured (Zikeli et al. 2002, 2005). Black carbon results in an increase total organic C and C/N ratio. Black carbon is recalcitrant and causes reduced microbial C and N. It may increase microbial respiration, N mineralization and nitrification, but tends to inhibit the activities of certain enzymes leading to reduced mineralization of litter, and the formation of abnormally thick humus layers (Klose et al. 2004; Koschke et al. 2011). Fly ash microspheres composed of magnetite can greatly enhance the magnetic susceptibility of urban soils (Gladysheva et al. 2007; Howard and Orlicki 2015). Howard et al. (2016) tested soil magnetic susceptibility as a proximal sensing method for mapping soils in the urbanized terrain of Detroit, Michigan. A map of magnetic susceptibility was constructed based on soils at abandoned, derelict residential home sites. The map revealed a striking positive magnetic anomaly centered over the heavily industrialized terrain in the southwestern part of the city (Fig. 10.14). This anomaly is thought to delineate the geographic extent of fly ash-impacted soils based on microscopic analysis showing the presence of magnetic microspheres.

Hiller (2000) studied anthropogenic soils at an abandoned railroad yard in Germany, much of which was built between 1907 and 1935. The soils were

10.6 Urban Soils Impacted by Coal-Related Wastes

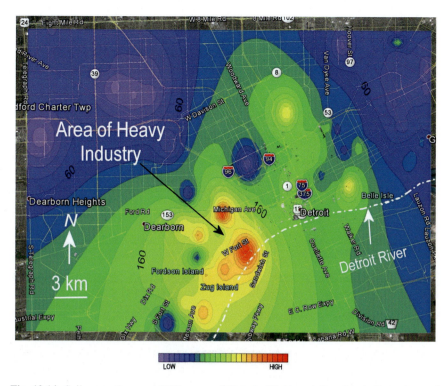

Fig. 10.14 Soil magnetic susceptibility map of Detroit, Michigan inferred to indicate the areal extent of fly-ash impacted soils. Highest values are in heavily industrialized area in southwestern part of the city. From Howard et al. (2016)

developed on raised land surfaces, 1.5–2 m higher than the adjacent natural landscape, that were underlain by anthropogenic deposits of coal ash, cinders, coked coal, unspent coal, and other industrial waste materials. These anthropogenic deposits contained an average of ∼58% coarse fragments, and had large voids (41% by volume) that were more or less filled with fine earth fraction. Compared to the natural soil, the anthropogenic soils were significantly lower in pH, CEC, and carbonate and nutrient content (Table 10.9). They were also characterized by a higher bulk density and zero water-holding capacity. Levels of total C, P and N were moderate to high, whereas plant-available levels were very low. This was attributed to the presence of black carbon in the form of coal and sooty coal ash. This is consistent with the fact that, despite their age, the anthrosoils had wide C/N ratios of 43–98, compared with 17–19 in the natural soil. The skeletal fabric of the railroad soils is reminiscent of that commonly observed in mine soils, as described in Chap. 9. This accounts for the low water-holding capacity, and suggests that the railroad soils have been strongly leached. Leaching probably accounts for the lower pH and lack of carbonate in the railroad soils, compared to the natural soils which contained about 10% carbonate. The railroad yard soils would be classified as

Table 10.9 Comparison of a natural soil and a railroad yard soil from Germany (Hiller 2000)

Horizon	Depth	pH	Organic C	CEC	Total P	Total N	Sand	Silt	Clay	Bulk density
	cm		%	cmol kg^{-1}	mg kg^{-1}	%	%	%	%	%
Natural soil (Calcic Cambisol)										
Ap	0–10	7.2	1.8	15.1	839	0.17	50	35	15	1.47
Ap2	10–20	7.3	1.5	14.5	761	0.15	50	35	15	1.50
Ap3	20–30	7.4	1.1	12.8	705	0.11	50	35	15	1.57
Bw	30–42	7.7	0.2	10.2	439	0.06	45	43	12	1.44
Railroad yard soil (Technosol)										
^A1	0–4	4.5	48.4	12.3	823	0.76	72	25	3	1.00
^A2	4–8	5.0	17.7	9.7	1480	0.27	82	18	0	1.60
^C1	8–16	5.7	25.1	11.3	716	0.38	74	21	5	1.50
^C2	16–48	6.0	32.0	15.3	761	0.57	73	24	3	1.60

Horizons modified from original description

skeletic Technosols according to the WRB, whereas natural soils in the same area are Cambisols (Inceptisols). A skeletic railroad yard soil is shown in Fig. 10.3d, and described in Table 10.4. It is characterized by a layer of densely packed, gravel-size coal cinders containing little fine earth material.

Coal tar is a black, sticky thermoplastic material produced along with ammonia when coal is carbonized to make coke, or when coal is gasified in manufactured gas operations. Coal tars and ammonia were important chemical feed stocks for chemical manufacturing industries, and used to make various products including artificial dye colors. Coal tar sludge (defined as a pumpable slurry with a solids content of 15–85%) from manufactured gas operations (Fig. 10.13b) were often stored in large underground tanks, and sold as a waterproofing material, or as a binder for road aggregates. Coal tar is a highly viscous, sticky, carbonaceous material comprised primarily of toxic polycyclic aromatic compounds. Coal tar produced as a by-product of coking and manufactured gas operations has adversely affected urban soils in many places. Coked coal was widely used in foundries to make cast iron, and was especially important in the manufacture of steel, where the carbon must be free of volatiles and ash. The coking process was also used to manufacture gas in so-called "town gashouses." In the gasification process, bituminous coal was heated in airtight ovens, and the gas driven off was then distributed through a system of pipes to nearby homes and commercial buildings. In the United States, manufactured gas was used for the first streetlights beginning in the mid-1800s, and the industry flourished until the 1880s when gas lights began to be replaced by electric lights powered by Edison Illuminating Company (founded in

10.6 Urban Soils Impacted by Coal-Related Wastes

1886). Nevertheless, manufactured gas-fueled stoves and furnaces were in great demand during the late 19th and early 20th centuries. The industry began to decline due to the increasing availability of natural methane gas via high-pressure pipelines around 1936, and the era of manufactured gas generally ended between about 1955 and 1975. Remediation of contaminated soils at former manufactured gas plants and town gashouse sites is difficult because of the materials handling problems created by the stickiness of coal tar.

10.7 Classification of Urban Soils

Most modern urban soils are classified as Technosols, according to the World Reference Base (IUSS Working Group 2015). These are basically soils that have $\geq 20\%$ (by volume) artifacts of technogenic origin in the upper 100 cm. Urban soils in residential areas are mainly Urbic Technosols (Table 10.10), which are characterized by an artifact assemblage indicative of human habitation. Some urban soils in modern residential areas may have a hortic horizon ≤ 50 cm thick, hence

Table 10.10 Basic types of urban soils and their classification according to the World Reference Base (IUSS Working Group 2015)

Reference Soil Group	Soil type	Geocultural setting	Description
Technosol	Urbic	Residential	Technogenic artifact assemblage comprised of $\geq 35\%$ (by volume) building rubble and artificial objects indicative of human habitation
	Spolic	Industrial	Technogenic artifact assemblage comprised of $\geq 35\%$ (by volume) industrial wastes including building rubble, mine spoil, dredgings, slag, cinders, or ash; includes hyperskeletic soils
	Garbic	Landfill	Technogenic artifact assemblage comprised of $\geq 35\%$ (by volume) organic wastes; includes soils with reductic and linic characteristics
	Ekranic	Manufactured land	Soils sealed by, or containing, one or more layers of technic hard material; includes certain soils with isolatic and leptic characteristics
	Isolatic	Rooftop	Soils in pots or rooftop gardens; also known as edifisols
Anthrosol	Hortic-like	Residential	Soils with a hortic horizon <50 cm thick
	"Necric"	Cemetery	Grave soils with a subsurface residual accumulation of organic matter derived from decomposed corpse, accompanied by artifacts indicative of human interment

See Table 10.2 for further information

they do not meet the requirements for classification as a Hortic Anthrosol. They probably will be classified as a natural soil with a hortic qualifier, or they could be described informally as a hortic-like anthrosol. Urban soils on industrial land are Spolic Technosols, which are characterized by an artifact assemblage comprised mainly of industrial wastes. Those with <20% (by volume) fine earth fraction may be classified as Hyperskeletic Technosols. Landfill soils characterized by artifacts in the form of organic wastes are Garbic Technosols. If they are reduced as a result of landfill gases they are Reductic Technosols, and those with a constructed geomembrane near the ground surface may be classified as a Linic Technosol. Areas of manufactured land, characterized by technic hard material, are classified as Ekranic and Leptic Technosols. They are known as ekranozems in the Russian classification (Prokofeva et al. (2001). Soils found in rooftop gardens or in pots are Isolatic Technosols (Edifisols of Charzynski et al. 2015). Grave soils in cemeteries are referred to informally here as "Necric" Anthrosols, as discussed in Chap. 8.

According to U.S. Soil Taxonomy (Soil Survey Staff 2014), urban soils formed in HTM are classified according to the definitions of a "buried soil" and a "surface mantle of new material." A buried soil is a sequence of one or more genetic horizons covered with a surface mantle of new material ≥ 50 cm thick. A surface mantle of new material is a layer of HTM which is unaltered in its lower part and is underlain by one or more buried genetic horizons. The mantle can have an epipedon, a cambic horizon, or both, but an unaltered layer (e.g., ^C horizon) ≥ 7.5 cm thick must be present in the basal part of the mantle. If the surface mantle of HTM is ≥ 50 cm thick and the lower part if unaltered, then the soil order is determined based on diagnostic characteristics of the mantle, and the surface of the mantle is taken as the starting depth for determining control sections. On the other hand, if the surface mantle of HTM is ≥ 50 cm thick and the lower part is altered, then the entire profile is used to determine the soil order. If the surface mantle of HTM is <50 cm thick, then only the characteristics of the buried soil are used for classification (Soil Survey Staff 2011). To illustrate, consider the three soils shown in Fig. 10.15. The first soil was formed in a surface mantle of HTM <50 cm thick, therefore the buried soil was used to classify the pedon (Fig. 10.15a). In contrast, the second soil was formed in a surface mantle of HTM ≥ 50 cm, and because the lower 7.5 cm is not altered and not part of any diagnostic horizon, the profile developed in HTM was used to classify the pedon (Fig. 10.15b). The third soil was formed in a surface mantle of HTM ≥ 50 cm thick, but the lower 7.5 cm is part of a ^Bw horizon (i.e. it is altered), hence the entire profile was used to classify the pedon (Fig. 10.15c). The drawback to this approach is that the anthropogenic nature of a soil is apparent at the family and higher levels of classification only if the surface mantle is used for classification (e.g., Anthroportic Udorthent). Otherwise, the soil will have the same name used for a natural soil (e.g., Typic Hapludoll), and its anthropogenic nature will be obscured, except perhaps at the level of a series or phase. A possible remedy would be to add the prefix "anthro" to the subgroup name, e.g., Anthrotypic Hapludoll.

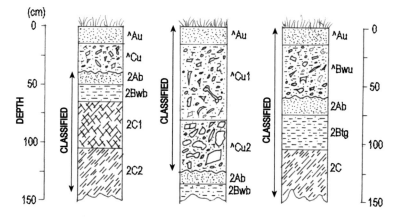

Fig. 10.15 USDA-NRCS method for classifying anthropogenic soils. Modified from Soil Survey Staff (2011)

References

Anderson JR, Hardy EE, Roach JT, Witmer RE (1976) A land use and land cover classification system for use with remote sensor data. US Geological Survey professional paper 964, 41 p

Arnold CL, Gibbons CJ (1996) Impervious surface coverage: the emergence of a key environmental indicator. Am Planners Assoc J 62:243–258

Bridges EM (1991) Waste materials in urban soils. In Bullock P, Gregory PJ (eds) Soils in the urban environment. Blackwell Scientific Publications, Oxford, England, pp 28–46

Burgess EW (1924) The growth of the city: an introduction to a research project. Pub Am Sociol Soc 18:85–97

Burghardt W (1994) Soils in urban and industrial environments. Z Pflanz Bodenk 157:205–214

Charzynski P, Hulisz P, Bednarek R, Piernik A, Winkler M, Chmurzynski M (2015) Edifisols—a new soil unit of technogenic soils. J Soils Seds 15:1675–1686

Chen G, Gan L, Wang S, Wu Y, Wan G (2001) A comparative study on the microbiological characteristics of soils under different land-use conditions from karst areas of southwest China. Chin J Geochem 20:558

Chen Y, Day SD, Wick AF, Strahm BD, Wiseman PE, Daniels WL (2013) Changes in soil carbon pools and microbial biomass from urban land development and subsequent post-development soil rehabilitation. Soil Biol Biochem 66:38–44

Craul PJ (1985) A description of urban soils and their desired characteristics. J Arboric 11:330–339

Cremeens D, Parobek J, Coyne J, Miller C, Dunham C, LaQuatra J (2012) Permanent vegetation establishment on manufactured soil at a former slag disposal pile in Pittsburgh, Pennsylvania: lessons learned. Soil Horizons, Madison, WI, 8 pp

Cuffney TF, Falcone JA (2008) Derivation of nationally consistent indices representing urban land intensity within and across nine metropolitan areas of the coterminous United States. USGS Sci Inv Rep 2008-5095

Cunningham MA, Synder E, Yonkin D, Ross M, Elsen T (2008) Accumulation of deicing salts in soils in an urban environment. Urban Ecosyst 11:17–31

Doerr SH, Shakesby RA, Walsh RPD (2000) Soil water repellency: its causes, characteristics and hydro-geomorphological significance. Earth-Sci Rev 51:33–65

Edwards WM, Shipitalo MJ, Traina SJ, Edwards CA, Owens LB (1992) Role of *Lumbricus terrestris* (L.) burrows on quality of infiltrating water. Soil Biol Biochem 24:1555–1561

Eldridge DJ, Greene RSB (1994) Microbiotic soil crusts: a review of their roles in soil and ecological processes in the rangelands of Australia. Aust J Soil Res 32:389–415

Etana A, Larbo M, Keller T, Arvidsson J, Schjonning P, Forkman J, Jarvis N (2013) Persistent subsoil compaction and its effects on preferential flow patterns in a loamy till soil. Geoderma 192:430–436

Ford LR (1996) A new and improved model of Latin American city structure. Geogr Rev 86:437–440

Gladysheva MA, Ivanov AV, Stroganova MN (2007) Detection of technogenically contaminated soil areas based on their magnetic susceptibility. Eurasian Soil Sci 40:215–222

Harris CD, Ullman EL (1945) The nature of cities. Ann Am Acad Polit Soc Sci 242:7–17

Hartmann P, Fleige H, Horn R (2010) Changes in soil physical properties of forest floor horizons due to long term deposition of lignite fly ash. J Soils Seds 10:231–239

Hazelton P, Murphy B (2011) Understanding soils in urban environments. CSIRO Publishing, VIC, Australia, 160 pp

Hiller DA (2000) Properties of Urbic Anthrosols from an abandoned shunting yard in the Ruhr area, Germany. Catena 39:245–266

Howard JL, Olszewska D (2011) Pedogenesis, geochemical forms of heavy metals, and artifact weathering in an urban soil chronosequence, Detroit, Michigan. Environ Pollut 159:754–761

Howard JL, Orlicki KM (2015) Effects of anthropogenic particles on the chemical and geophysical properties of urban soils, Detroit, Michigan. Soil Sci 180:154–166

Howard JL, Orlicki KM (2016) Composition, micromorphology and distribution of microartifacts in anthropogenic soils, Detroit, Michigan USA. Catena 138:38–51

Howard JL, Shuster WB (2015) Experimental order one soil survey of vacant urban land, Detroit, Michigan. Catena 126:220–230

Howard JL, Clawson CR, Daniels WL (2012) A comparison of mineralogical techniques and potassium adsorption isotherm analysis for relative dating and correlation of late Quaternary soil chronosequences. Geoderma 179–180:81–95

Howard JL, Dubay BR, Daniels WL (2013) Artifact weathering, anthropogenic microparticles, and lead contamination in urban soils at former demolition sites, Detroit, Michigan. Environ Pollut 179:1–12

Howard JL, Ryzewski K, Dubay BR, Killion TK (2015) Artifact preservation and post-depositional site-formation processes in an urban setting: a geoarchaeological study of a 19th century neighborhood in Detroit, Michigan. J Archaeol Sci 53:178–189

Howard JL, Orlicki KM, LeTarte SM (2016) Evaluation of some proximal sensing methods for mapping soils in urbanized terrain, Detroit, Michigan USA. Catena 143:145–158

Hoyt H (1939) The structure and growth of residential neighborhoods in American cities. Federal Housing Administration, Washington, DC

Huot H, Simonnot MO, Marion P, Yvon J, De Dontao P, Morel JL (2013) Characteristics and potential pedogenetic processes of a Technosol developing on iron industry deposits. J Soils Seds 13:555–568

Huot H, Simonnot MO, Morel JL (2015) Pedogenetic trends in soils formed in technogenic materials. Soil Sci 180:1–11

IUSS Working Group (2015) World Reference Base for Soil Resources 2014, update 2015 International soil classification system for naming soils and creating legends for soil maps. World Soil Resources Reports No. 106. FAO, Rome

Jim CY (1998) Urban soil characteristics and limitations for landscape planning in Hong Kong. Landscape Urban Plan 40:235–249

Kim DH, Shin MC, Choi HC, Seo CI, Baek K (2008) Removal mechanisms of copper using steel-making slag: adsorption and precipitation. Desalination 223: 283-289 Kim K, Asaoka S, Yamamoto T, Hayakawa S, Takeda K, Katayama M, Onoue T (2012) Mechanisms of hydrogen sulfide removal with steel making slag. Environ Sci Technol 46:10169–10174

Klose S, Wernecke KD, Makeschin F (2004) Microbial activities in forest soils exposed to chronic depositions from a lignite power plant. Soil Biol Biochem 36:1913–1923

References

Koschke L, Lorz C, Furst C, Glaser B, Makeschin F (2011) Black carbon in fly-ash influenced soils of the Dubener Heide region, central Germany. Water Air Soil Pollut 214:119–132

Meuser H (2010) Contaminated urban soils. Springer, Dordrecht, 320 pp

Mullins CE (1991) Physical properties of soils in urban environments. In: Bullock P, Gregory PJ (eds) Soils in the urban environment. Blackwell Scientific Publications, Oxford, England, pp 87–118

Muralha VSF, Rehren T, Clark RJH (2011) Characterization of an iron smelting slag from Zimbabwe by Raman microscopy and electron beam analysis. J Raman Spectrosc 42:2077–2084

Oh C, Rhee S, Oh M, Park J (2012) Removal characteristics of As (III) and As (IV) from acidic aqueous solution by steel making slag. J Hazard Mater 213–214:147–155

Pouyat RV, Yesilonis ID, Russell-Anelli J, Neerchal NK (2007) Soil chemical and physical properties that differentiate urban land-use and cover types. Soil Sci Soc Am J 71:1010–1019

Proctor DM, Fehling KA, Shay EC, Wittenborn JL, Green JJ, Avent C, Bigham RD, Connolly M, Lee B, Shepker TO, Zak MA (2000) Physical and chemical characteristics of blast furnace, basic oxygen furnace, and electric arc furnace steel industry slags. Environ Sci Technol 34:1576–1582

Prokofeva TV, Sedov SN, Stroganova MN, Kazdym AA (2001) An experience of the micromorphological diagnostics of urban soils. Eurasian Soil Sci 34:879–890

Puskas I, Farsang A (2009) Diagnostic indicators for characterizing urban soils of Szeged, Hungary. Geoderma 148:267–281

Santini TC, Fey MV, Smirk MN (2013) Evaluation of soil analytical methods for the characterization of alkaline Technosols: I. Moisture content, pH, and electrical conductivity. J Soils Seds 13:1141–1149

Scalenghe R, Ferraris S (2009) The first forty years of a Technosol. Pedosphere 19:40–52

Scharenbroch BC, Lloyd JE, Johnson-Maynard JL (2005) Distinguishing urban soils with physical, chemical, and biological properties. Pedobiologia 49:283–296

Schmidt MWI, Knicker H, Hatcher PG, Kogel-Knabner I (2000) Airborne contamination of forest soils by carbonaceous particles from industrial coal processing. J Environ Qual 29:768–777

Sere G, Schwartz C, Ouvrard S, Renat JC, Watteau F, Villemin G, Morel JL (2010) Early pedogenic evolution of constructed Technosols. J Soils Seds 10:1246–1254

Shaw P, Reeve N (2008) Influence of a parking area on soils and vegetation in an urban nature reserve. Urban Ecosyst 11:107–120

Short JR, Fanning DS, McIntosh MS, Foss JE, Patterson JC (1986a) Soils of the Mall in Washington, DC: I. Statistical summary of properties. Soil Sci Soc Am J 50:699–705

Short JR, Fanning DS, McIntosh MS, Foss JE, Patterson JC (1986b) Soils of the Mall in Washington, DC: II. Genesis classification and mapping. Soil Sci Soc Am J 50:705–710

Shuster WD, Dadio S, Drohan P, Losco R, Shaffer J (2014) Residential demolition and its impact on vacant lot hydrology: implications for the management of stormwater and sewer system overflows. Land Urban Plan 125:48–56

Soil Survey Staff (2000) Urban soil compaction. Soil quality urban technical note no. 2, USDA, NRCS, 4 pp

Soil Survey Staff (2005) New York City reconnaissance soil survey. United States Department of Agriculture, Natural Resources Conservation Service, Staten Island, NY, 52 pp

Soil Survey Staff (2011) Buried soils and their effect on taxonomic classification. Soil survey technical note no. 10, USDA, NRCS, 7 pp

Soil Survey Staff (2014) Keys to soil taxonomy (12th edn). US Department of Agriculture, Natural Resources Conservation Service, 372 pp

Tasakiridis PE, Papadimitriou GD, Tsivilis S, Koroneos C (2008) Utilization of steel slag for Portland cement clinker production. J Hazard Mater 152:805–811

Vance JE Jr (1964a) Geography and urban evolution in the San Francisco Bay area. University of California, Institute of Governmental Studies, Berkeley, CA

Vance JE Jr (1964b) The continuing city: urban morphology in western civilization. Johns Hopkins University Press, Baltimore, MD, 534 pp

Vance JE Jr (1990) The continuing city: Urban morphology in western civilization. Baltimore, MD, Johns Hopkins University Press Vauramo S, Setala H (2011) Decomposition of labile and recalcitrant litter types under different plant communities in urban soils. Urban Ecosyst 14:59–70

Wang M, Markert B, Shen W, Chen W, Peng C, Ouyang Z (2011) Microbial biomass carbon and enzyme activities of urban soils in Beijing. Environ Sci Pollut Res 18:958–967

Yang Y, Paterson E, Campbell CD (2001) Urban soil microbial features and their environmental significance as exemplified by Aberdeen City, UK. Chin J Geochem 20:34–44

Yildirim IZ, Prezzi M (2011) Chemical, mineralogical, and morphological properties of steel slag. Adv Civil Eng. Article 463638, 13 pp

Yunusa IAM, Manoharan V, Odch IOA, Shrestha S, Skilbeck CG, Eamus D (2011) Structural and hydrological alterations of soil due to addition of coal fly ash. J Soils Seds 11:423–431

Zhao D, Yang FL, Wang R, Song Y, Tao Y (2013) The influence of different types of urban land use on soil microbial biomass and functional diversity in Beijing, China. Soil Use Manag 29:230–239

Zikeli S, Jahn R, Kastler M (2002) Initial soil development in lignite ash landfills and settling ponds in Saxony-Anhalt, Germany. J Plant Nutr Soil Sci 165:530–536

Zikeli S, Kastler M, Jahn R (2005) Classification of anthrosols with vitric/andic properties derived from lignite ash. Geoderma 124:253–265

Chapter 11
Epilogue

It seems clear that even after anthropogenic activities have ceased, the impacts of humans on soils can persist for hundreds and perhaps thousands of years. This is shown by the fact that ancient Hortic Anthrosols (e.g., garden and midden soils) and Pretic Anthrosols (i.e. Amazonian Dark Earth soils) have strongly influenced the colonization pattern of modern native plant species and crop production. On the other hand, soils created by human disturbance can be quite resilient. For example, certain urban and mine-related anthropogenic soils have rebounded to a quasi-natural state in ~60 years even without the application of reclamation methods. The persistent impacts of human activities on soils have had both positive and negative consequences. Soils in ancient agricultural settings that were farmed for ≤ 500 years showed tell-tale signs of degradation, such as decreased organic matter, N and P content, which have persisted for at least 800 years. In contrast, soils farmed for ≥ 1000 years showed beneficial effects such as higher fertility in moist climates, or lower salinity in dry climates, that have persisted for 500 years or more. These benefits are reaped by those who currently farm ancient agricultural soils, and may provide useful information leading to future methods of sustainable agriculture. The natural resilience of anthropogenic soils is similarly encouraging from the standpoint of urban and mined land reclamation and revitalization.

The presence of an abnormally thick, black, organic carbon-rich A horizon is perhaps the most universal characteristic that distinguishes anthropogenic soils from those of natural origin. The black color is the result of both highly recalcitrant humic substances, and microartifacts comprised of bitumen or black carbon. Studies of garden soils (Hortic Anthrosol), mine soils (Spolic Technosol) and urban soils (Urbic Technosol) show that anthropogenic A horizons at least a few centimeters thick can develop rapidly, typically in less than 25 years, and sometimes in less than 5–10 years. Anthropogenic A horizons can grow to be 25–30 cm thick after 100 years, and 100–200 cm thick after 1000 years or more, of pedogenesis under moist and well-vegetated conditions. B horizon genesis is slower and irregular, at least for the first ~100 years of soil formation, although proto-cambic horizons are occasionally reported in urban and mine soils after only 20–30 years. Cemetery soils are also reported to have weakly developed cambic or cambic-like horizons after 112–115 years of soil formation. Urban and mine soils commonly showed evidence of accelerated weathering during the early stages (6 months–10 years) of pedogenesis as a result of human activities. This includes rapid physical disintegration of coarse fragments, chemical weathering of certain artifacts, and leaching of soluble salts and carbonates. In contrast, studies of soils on ancient burial mounds showed that well-developed anthropogenic cambic horizons required 500–2000 years to form, and argillic horizons required 2500–3000 years to form, after human habitation had ceased.

The presence of artifacts and/or microartifacts is also a signature feature of anthropogenic soils. These include artificial objects which have been manufactured or altered by human action, as well as natural materials (e.g., coal) which are only present at a given locality because of human activities. Artifacts are typically household items, tools, food-related wastes, wood and coal combustion products, waste building materials, etc. They are often comprised of mineraloids, minerals, or mineral assemblages not found in nature. The impacts of these materials, and their weathering products, on anthropogenic soils and the environment are poorly understood and deserve further study. Microartifacts are important because even when present in small quantities (<10% by weight) they can have a significant effect on the physical, chemical and geophysical properties of anthrosoils. The presence of artifactual materials also has stratigraphic significance, and serves as an excellent indicator of the Anthropocene Series in the geologic record.

Human-modified soils often require reclassification as a different type of soil. For example, certain shell-midden and midden-mound soils were classified as Alfisols, whereas local natural soils were Ultisols. Anthropogenic soils formed in human-transported material were perhaps most commonly classified as Entisols, but some were classified as Inceptisols, Alfisols and even Ultisols, according to U.S. Soil Taxonomy. Anthropogenic soils in urban and mine-related geocultural settings were usually classified as Technosols, according to the World Reference Base, whereas Anthrosols generally predominated in agricultural and archaeological settings. Several possible problems regarding anthropogenic soils that remain to be addressed by these soil classification schemes include: (1) the effects of bitumen and pyrogenic black carbon on differentiate involving organic carbon content, (2) the definition of the necric horizon and classification of grave soils, and (3) the identification and use of microartifacts at lower taxonomic levels of classification.

Although anthropogenic soils in agricultural settings are by far the most significant on the basis of total land area coverage, mine soils on artificial terraces created by strip mining can be locally important, especially in areas of rugged terrain where the natural landscape has a very limited amount flat land suitable for agriculture or habitation. It is also the anthropogenic soils of urban areas which are most commonly encountered by the majority of the population. Given that the original site chosen for the founding of most cities was almost universally based on natural resources, especially soil fertility, urbanization typically occurs today at the expense of prime agricultural land. Thus, there is an increasing demand for better knowledge about the impacts of human activities on the landscape in general, and

11 Epilogue

about the characteristics of anthropogenic soils in particular. The amount of urban land is increasing worldwide at a rapid rate driven by the exponential growth in global population, and the progressive shift of populations from rural to urban areas. Thus, it is anticipated that the land area covered by anthropogenic soils will continue to grow at a dramatic rate, and the need for better scientific information regarding anthropogenic soils is expected to increase well into the foreseeable future.

Printed in the United States
By Bookmasters